Stanier '8F' No 48606 residing outside Burton Shed.

BRITISH STEAM NOSTALGIA

BRITISH STEAM NOSTALGIA

COLIN GARRATT

Patrick Stephens
Wellingborough, Northamptonshire

First published in 1987

British Library Cataloguing in Publication Data

Garrat, Colin
British steam nostalgia.
1. Locomotives—Great Britain—History
—20th century—Pictorial works
I. Title
625.2'61'0941   TJ603.4.G7

ISBN 0-85059-863-X

Title page *See page 92.*

*Patrick Stephens Limited is part of
the Thorsons Publishing Group*

Printed in Great Britain by
Butler & Tanner Ltd, Frome and London

Contents

Introduction   6

Chapter 1   **Great Western**   8

Chapter 2   **London Midland & Scottish**   26

Chapter 3   **Industrials**   68

Chapter 4   **British Railways**   88

Chapter 5   **Southern**   98

Chapter 6   **London & North Eastern**   114

Index   142

Introduction

A book critic recently accused me of being a 'purveyor of steam nostalgia'. He was appalled that anyone could wax lyrical about Britain's 'dirty, inefficient, uneconomic and out-of-date railway system' as it was during the decade following World War 2.

Nostalgic I remain — and unashamedly so, for I believe in the steam locomotive's animated simplicity. I believe that a comprehensive railway network in prime service to the nation is the most civilized form of land transport we have yet devised and I also believe that variety and diversity is the spice of life. Britain's railways had all of these attributes.

But why trainspotting? Because it reflected the

reverence we felt for the wonders of the railway. Every other boy was a loco-spotter and during evenings, weekends and holidays the well known spotting places throughout Britain attracted vast numbers of people, whilst there was always companionship at one's local bridge or cutting. Grass was worn off the embankments and we sat on hard dusty patches; cycles, pop bottles, sandwiches, note books and ABCs were the vital elements to one of the finest dramas of all time. Britain's 30,000 steam locomotives — embracing 600 different types — seeped into every industrial nook and cranny. Whoever understood railways in 1950 had his finger on the nation's pulse, for railways were the economic heartbeat of the nation.

And the magic was going to last forever — steam railways were permanent institutions. We had heard about a government plan to phase them out, but it couldn't happen in our lifetime.__

But in a few tragic years the magic was dissipated and the most awe-inspiring of man's creations vanished, taking with it half the railway network — fifty per cent of our greatest industry — as the motorway network spread across Britain like a cancer: motor cars began to settle like flies in every open space and juggernauts pervaded the land in a free-for-all undisciplined, anti-social and wasteful transport system which sucked the energy and vitality from our railways and turned the industry into a disillusioned travesty of everything it once stood for. Britain's railways had provided a comprehensive, safe, energy-efficient and environmentally kind transport system: now they were decimated and those with vested interests in roads were exceedingly pleased.

The spirit of trainspotting on an intensely used system might be compared with current events at Cley marshes in Norfolk, one of Britain's finest birdwatching places with a recorded total of 334 species. Apart from many resident species, Cley attracts innumerable migrants on spring and autumn movements, including rarities of North American and East European origin.

Nancy's Cafe in Cley has become one of several unofficial information centres for sightings of rare birds throughout the country. The telephone rings constantly and the diary, randomly filled in by birdwatchers — or 'twitchers' — constitutes an excellent daily guide to interesting events. The analogy with trainspotting is remarkable, as these diary extracts show:

'Nancy's
Monday 30 June 1986
Phone call from Hickling. The 'Broad Billed Sandpiper' [a rare North American wader] resident here for past three days, flew off north accompanied by two 'Dunlins' at 10.00 this morning; watch for their possible arrival at Cley.'

'Derby trip
Saturday 24 July 1954
Ex-works Swansea 'IF' Tank No 41769 which has remained here on yard shunting over the last ten days was seen heading light towards Burton-Upon-Trent around 11.00 this morning — presumably on its journey home.'

Twitchers are not ornithologists — any more than trainspotters are engineers — but both have a superb knowledge of their subject. This knowledge needs to be utilized beyond mere enthusiam. The ongoing destruction of our railways is justified on 'economic grounds'; now the same specious argument is being put forward for the widespread destruction of Britain's natural environment with the inevitable diminution of our fauna and flora.

The fraternity has to be involved. Whatever level our interest we are custodians of the heritage in our time and it is our duty to play a positive role in protecting that heritage. Britain's railway needs all the support it can get from the millions of people who supposedly care about it and — as everyone knows — the same is true of our threatened wildlife.

Newton Harcourt days — Colin Garratt at the age of 10 **above** *and today by his childhood bridge on the Midland main line* **opposite**.

Recently the television news carried pictures of BR's Advanced Passenger Train being broken up in a Rotherham scrapyard. I was as distressed to see this as I had been twenty years ago when I witnessed 'Jubilees' being scrapped in the same area. 'Jubilees' or APT it mattered not; both were part of the railway and both had a future.

Yes, I am unashamedly nostalgic; but only because I am prepared to fight for railways past, present and future. Nostalgia ought not to be an indulgence, but part of an ongoing creative process; it is in this spirit that I offer the following pages.

Colin Garratt
Newton Harcourt Leicestershire
February 1987.

Chapter 1

Great Western
Tanks — Branch and suburban

A properly structured railway consists of a diversity of services, ranging from heavy, long distance traffic on main lines down to the lowliest of branch line duties. The '1400's belong to this last category and, though humble, they formed part of a network of feeder and secondary lines without which the main line systems are inevitably devitalized and the railway unable to play a properly co-ordinated role in service to the

These delightful branch tanks — built under Collett between 1932/6 — were for the lightest work. Though new, they replaced almost identical engines of nineteenth century origin.

Pure Great Western: '4100' Class suburban 2-6-2T, No 4179 in repose, alongside a typical Great Western coaling stage.

nation.

Until the mid-1950s — when the nation's railway remained virtually complete — the '1400's were frequently seen in bay-platforms at busy stations with one or two chocolate or cream-coloured coaches waiting for connection with incoming services before chugging away to some delightful rural retreat. The system was economical, efficient and reliable. The class consisted of 95 engines and most were fitted with push and pull apparatus for rail motor work. They were extremely lively and one was recorded running at almost 80 mph whilst pushing a coach between Stonehouse and Gloucester.

The closure of branch lines and the influx of diesel railcars began to render the '1400's redundant, and withdrawals began in 1956, but a few survived until the mid-'60s.

The Great Western practised a remarkable continuity of locomotive design, and as the '1400's were an update of a nineteenth century type, the '4100' Class suburbans were direct descendants of a Churchward design of 1903. No 4179 typified the style of suburban/local stopping tank which was prolific over the Great Western network for more than half a century — albeit that she wasn't built until 1949, after the formation of British Railways.

'Castles' — Ignominous end

One of the last surviving 'Castles' ekes out its existence at Oxley depot, Wolverhampton.

It was hard to believe that the 'Castles' would ever be withdrawn. Watching them as a young boy, they seemed so superbly modern; so clean and well-maintained, and complete masters of their work. Few other classes inspired such confidence.

I first saw them on an early visit to Birmingham's Snow Hill, and gazed with wonderment at these Great Western beauties with their copper embellishments, sensuous curves and unashamedly bold names. By the time I was twelve, I could recount *Treganna Castle's* exploits on the Cheltenham Flyer in 1932, when she ran the 77 miles from Paddington to Swindon in 56 minutes — an average speed of 82 miles an hour, start

to stop. They were true greyhounds, and never will I forget seeing No 5054, *Earl of Ducie* passing through Cholsey at 100 miles an hour.

The run-down of steam during the 1960s was obviously a deeply depressing time, but the withdrawal of the indestructible 'Castles' — mainstay of the Great Western's express passenger fleet for over thirty years — was especially tragic, and seeing sights like those on this page caused depression which lasted for hours afterwards, and I remember that we were on the point of not going to Oxley depot again, as we simply could not bear to see what was happening to these glorious engines.

The beauty of the 'Castle' design is evidenced in these studies of condemned examples awaiting scrapping in South Wales in 1964.

'Castles' — Revival

Above *We arrived at the Broughton Road crossing on the Market Harborough to
Northampton line with only minutes to spare, having raced the train by road
from Newton Harcourt.*
Above left *Nottingham Midland Shed at lunchtime with* Clun Castle
on the ashpits.
Left *Evening's shadows lengthen as* Clun Castle *heads away from
Northampton station.*

Clun Castle was the hero of steam's lost cause in the
mid-'60s. How she contrasted with her begrimed and
work-weary compatriots as, reserved for special runs,
she operated innumerable rail tours! These were the
days when Britain's surviving steam engines were
begrimed, shorn of all adornments and ekeing out a
living death. Millions flocked to the tracksides to see
Clun Castle, whether she was heading the last steam
train from Paddington to Chester, or even the last
steam train from Paddington itself.

In March 1965, *Clun Castle* worked an Ian Allan
rail tour through the Midlands. It was a day of
continuous sunshine, and we followed her from Derby
to Nottingham. From here, she was scheduled to work
south along the Midland main line to Market Harborough, before taking the now closed line to Northampton. Having photographed her on the coal road at
Nottingham shed, we drove to my childhood bridge
at Newton Harcourt, and the picture we made is on
page 29. Now came the most difficult part, as we had
to race the train from Newton Harcourt to Broughton
crossing, north of Northampton. We were travelling
in a Sunbeam Talbot, and my friend drove fast and
confidently, but we heard *'Clun'* whistle as we arrived
and just had time to make a picture (above) whilst
another mad chase, this time into Northampton,
caught the glorious engine leaving under the electrification catenary.

Panniers

For a young boy accustomed to seeing shunting tanks with traditional flat sides, the once-ubiquitous panniers of the Great Western came as a great surprise. It seemed as if there was one in every yard — often in quite small stations — a reminder of the tremendous utilization of our railways a mere thirty years ago. The vastness of the Great Western system meant that the panniers were amongst the hardest of all classes to clear, and on train journeys through Great Western territory one was on constant tenterhooks least a pannier was missed at some unexpected location. We invariably saw them, but often just the chimney and dome protruding above wagons.

The Great Western began to utilize panniers in the early years of this century, initially by re-building earlier saddle tanks, and it was this conversion which created the precedent for the '5700' Class which first appeared in 1929. Over the following twenty years, the '5700's reached a total of 863 engines whilst, in all, the Great Western had about a thousand panniers in service. They were found on a wide range of light work from shunting and tripping to branch passenger.

Right *One of my best finds of the '50s was No 2138, a member of the '2021' Class. These engines were rebuilt from a class of saddle tanks introduced in 1875 — I caught her ending her days shunting at Hereford. All had gone by 1959.*

Below right *Work-weary '5700' Class pannier tank, No 3620, in begrimed condition and with number plate missing, on shunting duty in the mid-'60s.*

Below *No 4680 was a member of the famous '5700' Class of pannier tanks, introduced by Collett in 1929.*

'Halls'

Above *'Hall' Class 4-6-0, No 6974*, Bryngwyn Hall.
Below *GW 'Hall' Class 4-6-0, No 6991*, Action Burnell Hall.

The mystique of the 'Halls' was second only to that of the LMS 'Jubilees'. They were the most numerous-named class on British Railways, with over 330 in service, and their names — infinitely less diverse than the 'Jubilees', but never boring — conjured up images of a vast world beyond our immediate consciousness. Where were these historic 'Halls', many of which one couldn't even pronounce; what was their history? Who lived in them? Yes, they came from a dream world which, having been glimpsed, heightened our childhood sense of wonderment and insight into the monumentality of history and creation — a sense of wonderment which the most privileged of us never lose.

The 'Halls' were spread throughout the entire Great Western network and provided the very essence of exciting trainspotting as, being mixed traffic engines with a long-distance running capability, rare birds often turned up with special workings which took them miles off their own territory.

My first experience of the 'Halls' was in 1951, when an older spotting friend took me to Birmingham Snow Hill Station for the day. I was eleven, and the purpose of the trip was to try to see one of the last 'Bulldogs'. I remember leaving Oadby at seven in the morning to catch our train to Birmingham, and my sense of wonder upon crossing to Snow Hill to see for the first time the magic of the Great Western; shafts of smoky sunlight poured into that magnificently busy station and fell onto the green engines to heighten their burnished copper and brass work. We didn't see our 'Bulldog' — in fact I never saw one — but we did see No 2934, *Ashton Court,* my only 'Saint', and the design from which the 'Halls' were derived. But throughout that day at Snow Hill we saw lots of 'Halls': 4908, *Broome Hall;* 4949, *Packwood Hall;* 4983, *Albert Hall;* 4926, *Holkham Hall,* and many more, but during the long spotting years which were to follow, there were many I didn't see. In fact, few of us got them all, apart from the most dedicated Great Western followers. I suspect one had to live deep within Great Western territory to achieve such distinction.

Below *GW 'Hall' Class 4-6-0, No 6998,* Burton Agnes Hall.

'Halls' at Banbury

Right *Classic Great Western semaphores at Banbury station.*

Above left and left *'Hall' Class 4-6-0, No 6930,* Aldersey Hall, *languishing on the condemned line at Banbury.*

The 'Halls' often ranged beyond their home territory and for us in Leicester there was a daily Bournemouth to York working which brought a 'Hall' into the Great Central every afternoon. The engine worked back to Banbury later that evening with a fish train.

We regularly cycled across to the Great Central to see the 'Western' as we called it. Usually the engine would be standing near the Fish Dock — an inaccessible part of the station area and on one occasion our rampant trespass there in pursuit of the copper-capped beauty led to an arrest by the railway police. False addresses were out of the question as we were in school uniform.

This time we didn't get away with it and we duly appeared before the Leicester Junior Magistrates Court. A uniformed officer solemnly read out the charge and we were each fined fifteen shillings and instructed never to venture onto railway property again!

I saw my last 'Halls' at Banbury, but depicting them here, in this way, is a travesty. Monochrome film does little justice to Great Western engines, whilst revealing them in their final hours makes a bitter contrast with the pride, cleanliness and performance which made the GWR the miracle it was.

The '28XX's

The long, slow moving freight train with a panting 0-6-0 at its head was becoming increasingly troublesome at the turn of the century. Busy lines were becoming cluttered as rail traffic continued to increase. The 0-8-0 which first appeared in 1889 was a natural progression, but it was unsuitable for the high speeds which were becoming necessary. But when Churchward introduced his 2-8-0 heavy mineral engine on to the Great Western in 1903, it emerged a lean, handsome beast which was years ahead of its time. It was considerably more powerful than anything used before and the leading wheels provided the necessary stability for faster running. The '28XX's introduction set the scene for the development of Britain's freight locomotives until the end of steam as a policy of regular medium-sized freight hauls demanded nothing larger than the 2-8-0 and it

was not until the eleventh hour of steam that the 2-8-0's supremacy was usurped by the British Railway's '9F' 2-10-0. Despite the '28XX's success it took the other railways a long time to catch up, so potent a workhorse was she, and in the case of the LMS, not until the '30s when Stanier introduced his '8F' — 32 years later!

After its initial trial, the first '28XX' was joined in 1905 by a further twenty engines and, by 1919, a total of 84 had been introduced. Twenty years later, more were built under Collett's regime and, by 1942, the class totalled 167 engines. More may well have been produced had not the government intervened by ordering a batch of wartime Stanier '8F' 2-8-0s from Swindon in 1942. These later '28XX's were identical to the original Churchward engines, apart from the provision of side-window cabs and outside steam

Above *The '28XX's delightful design is epitomized in this study of No 3844 being serviced at a typical Great Western coal stage.*

Above left *The outside steam pipes of No 3864 indicate that she belongs to the wartime batch built at Swindon between 1938/42.*

pipes. This extended building period reveals once again the GW's policy of producing proven designs over extremely long periods.

Scrapping commenced in 1958, when the original engine of 1903 was withdrawn, but survivors remained until the end of Great Western steam.

South Wales 1955

Above *A moment of triumph at Cardiff East Dock depot (88B) as I pose proudly on former Rhymney Railway 'R1' Class 0-6-2T No 41. She was especially interesting in that she retained her original domed boiler where as many were rebuilt with GW taper boilers.*
To Midland teenagers in the '50s, the South Wales valleys were a world away, and the remaining few survivors of such fabled railways as the Rhymney were hard to find in their shunting trips and valley haunts. All had gone by 1958.

Below left and below right *Another priority of this tour was to see the last engines from the Taff Vale railway. Many were remnants of their once-numerous 'A' Class 0-6-2Ts built between 1914 and 1921. Here, we see No 381 at Cardiff Cathays depot (88A) but, unlike the RR example above, she has been fitted with a Great Western boiler.*

One of the most memorable tours. For years we had dreamed of the old and rare classes from the South Wales companies; some well-known, like the Taff Vale and Rhymney, others obscure and abounding in fascination, like the Alexandra Docks or the Port Talbot railway. In the event, our tour coincided with the national rail strike of 1955 which meant a lot of extra walking to and from bus routes and I remember how blistered our feet were at the end of that week. However, the sheds were full of engines and although police were on duty at most depots, we were invariably able to negotiate admittance, not least since the danger element of moving engines had been removed. It was a hot, glorious, sunny week of endless depots and rare types — as happy a week as I have ever spent.

Below left and below right *A highspot of that week was our visit to Llanelly ('87F') where we were eager to see the former Bury Port and Gwendraeth Valley types which included 0-6-0Ts of several designs. On shed we found* Gwendraeth, *an 0-6-0ST of 1906, along with No 2198, a handsome 0-6-0T of typical Hudswell Clarke design, and one which was later developed by the company for general use — as epitomized by 'Tank Engine Thomas'. The foreman was extremely friendly to us, and arranged for No 2198 to be hauled outside by a pannier tank for these photographs — providing we took a picture of him too!*
Over thirty years later, at an audio visual show in nearby Gorseinon, I met an old Llanelly driver who, for a time, had BPVS's Kidwelly as his regular engine. When I recalled our visit and our friend, 'Mr James', he replied, 'the best forman we had and a great railway man — he died only recently'.
No 2198 was the last BPGV engine to remain in service, surviving until 1959.

Banbury 1966

Above *An unidentified Hall in the process of being cut up in the depot yard.*

Above right *Judy Maddock.*

Right *All that was left of an unidentified Grange 4-6-0 in the scrapyard at Banbury shed.*

Just as the decline of steam reached its peak in 1965, I abandoned my job to try to make a break as a professional jazz musician, and in the autumn of that year left the Midlands for London, accompanied by Judy Maddock who had sung so superbly with Colin Garratt's Superior Jazz Band in Leicester. We lived in Clapham Common and quickly formed a band, but within months it became evident that we'd never earn a living from the music and dejectedly returned to Leicester in the spring of 1966.

Steam was disappearing with ever-increasing rapidity and I realized that during those years of intense pre-occupation with music, innumerable photographic opportunities had been missed. From then on, until the end of BR steam in 1968, I tried to make up for lost time. We made many forays to Banbury where the last Great Western types could be found albeit heavily interspersed with ex-LMS Stanier types and BR standard '9F' 2-10-0s.

I was invariably with Judy Maddock, and if the pair of us sought solace in the railway to assuage our disappointments musically, it was never found, for the accompanying pictures sum up Banbury in its last months, the only bright spot being a resplendent *Clun Castle* allocated there and retained for working specials.

Chapter 2

London Midland & Scottish
My first pictures

Ex-Midland Railway '2P' 4-4-0, No 40364 — a Burton engine — waits to depart with a stopping train to Trent. Note the typical group of spotters.

Trafford Park 'Jubilee' No 45622, Nyasaland, piloted by ex-MR '2P' 4-4-0 No 40416 — a Derby engine — on a St Pancras to Manchester express.

These were my first railway photographs. It was a summer afternoon in 1951; I was eleven and asked my parents if I could borrow their box camera — I even recall their consternation that I might break it. But I can't remember what motivated me to take the photographs; it was a routine afternoon's spotting at the north end of Platform 2 at Leicester London Road Station. The thrill of trainspotting captivated our attention virtually to the exclusion of everything else — including schoolwork — and photography was little thought of.

And they are trainspotter's pictures; notice that I simply took the pictures as the trains came to a stop at the north end of the platform, which is where we sat for the afternoon. Yet the pictures were reasonable — as a first attempt and even my parents were surprised. The negatives of that first afternoon have long since been lost and all that remains are these prints, along with the one on page 39. Unbeknown, the seeds of my future career had been sown.

Brand new BR standard '5', No 73018 — Derby built, 1951 — waits to leave at 15.30 with a St Pancras to Nottingham express.

Leicester's own Ivatt 2-6-2T, No 41268, shunted Fox Street sidings for most of her working life, having replaced our two former Midland Railway '1P' 0-4-4Ts Nos 58072/3.

Newton Harcourt

After school one afternoon, my friend Roger Guillain suggested a cycle ride, and we rode from Oadby, over the fields to the tiny village of Newton Harcourt, where we paused on a Victorian railway bridge overlooking the main line, seven miles south of Leicester. Standing on our crossbars, looking down at the shiny metals, I willed a train to pass. Eventually a southbound coal train came into view; the engine was working hard and pumping grey smoke into the air. It was a Stanier '8F', and as the engine passed beneath, a mighty cloud of smoke caused us to jump from our crossbars — my first initiation of that most exotic of perfumes. We ran to the other parapet, thrilled by how the bridge shook as the wagons of shiny black coal slid below and hoped that the engine would reach the distant bridge before the last wagon passed below us — it did!

I was back at that enchanted bridge the next day, and indeed much of my remaining childhood was spent there, little dreaming that one day I would — from that very village — scour the earth to document such sights.

Left *The bridge from which I watched my first train in 1949.*

Below left *Ex-LMS Black '5', No 45254 heads a mixed freight beneath the childhood bridge and heads south towards Newton Harcourt village bridge, one quarter mile distant.*

Below *During our day with* Clun Castle *— described on page 13 — we left Nottingham to catch her passing beneath the bridge at Newton Harcourt.*

'Black 5's: 1

The 'Black 5's were probably the best all round steam design we produced. No other class achieved such widespread popularity. Their introduction revolutionised motive power on the LMS and caused the relatively early retirement of many highly distinctive pre-grouping types — not least former LNW engines. There were 842 'Black 5's allocated throughout the country, and they could be entrusted to virtually any task set before them. I have never known any engine man to speak badly of them and throughout my years of watching trains they always seemed masters of their work. Building continued from their inception in 1934 until 1950 after nationalisation and their reliability and ease of maintenance provided a mainstay of LMS motive power. So good were they that I doubt Stanier need have designed his 'Jubilees', for there was little a 'Jubilee' could do that a 'Black 5' couldn't. Certainly on the Midland main line in the '50s, the top expresses, though rostered for a 'Jubilee', would often be 'Black 5's, including the Thames-Clyde Express, and there never seemed any difference in performance or time-keeping. Fortunately, this was not realized in 1935, or we would have been deprived of the awe-inspiring 'Jubilees'.

Perhaps the same could be said of the 172 BR 'Standard 5's, introduced in 1951, and based on the 'Black 5'. Many people said at the time of the 'Standard's inception that 'Black 5's should be perpetuated instead and, good as the 'Standard 5's were — in retrospect, and given their short life — Black 5s would probably have proved more economical.

For trainspotters, the 'Black 5' represented as great a challenge as any. They were spread throughout the LMS network — and often beyond — from depots as far apart as Inverness and Bristol. There were a few vari-

30

ations within their ranks, the most interesting being numbers 44738-57 which were fitted with Caprotti valve gear. Based at Llandudno Junction, Longsight, Leeds and Bristol these engines were characterized by their huge outside steam pipes and lower running plates complete with splashers. They were a great joy to see and whenever we saw one approaching we hoped it would be one of the rarer Llandudno Junction batch. When I began spotting in 1949, the 'Black 5's in general were allocated to over seventy different depots. I remember odd spotters who did clear them, but few of us achieved such distinction, and I ended up needing 32. Not surprisingly, some 'Black 5's lasted until the very end of BR steam in August 1968.

Below left *Soft evening sunlight heightens the fine features of a Stanier 'Black 5' in repose on Banbury Shed.*

Below *A 'Black 5' on the turntable at Banbury, where these engines had replaced most of the traditional GW types there. The 'Black 5's were built over a sixteen year period, some after the formation of British Railways. No 44661 was one such engine, having been completed at Crewe in 1949.*

A Caley tank

Above *Former Caledonian Railway 0-6-0 dock tank, No 56159, one of six allocated to Glasgow, Polmadie. Notice the sister engine behind, with stove-pipe chimney.*

Right *Countless visits were made to Rose Grove depot at Burnley, and over those final years friendships were formed. Here is Fireman Walker at the regulator of Stanier '8F' No 48257.*

A picture taken thirty years ago, at Polmadie shed in Glasgow, of a fine little dock tank. Elusive — seldom seen and never at home — engines, whose quiet weekends were spent snuggled away in some remote sub-shed, hidden from view or knowledge, except by those with whom the engine's working lives were inextricably bound.

The class numbered 23 engines introduced by McIntosh in 1911, and, considering the once in-

credible maze of sidings, yards, factories and dock premises in the old industrial areas of the Clyde, I did well to see most of them, especially as some of the harder ones were tripping from such sheds as Daws-holme (65D), Kipps (65E), Grangemouth (65F) and Yoker (65G). Throughout our Scottish tours of 1955/6/7, the class remained intact, but withdrawal began in 1959 and in a short space of time — by 1961 — they were extinct.

Left *Spotters throughout England dreamed of seeing the four 'Mickey Namers', the four named examples of the 842-strong 'Black 5's. They were Scottish and rare, but transfers during the '60s brought No 45156, Ayrshire Yeomanry, to England; she is depicted here on Patricroft shed on Sunday 19 May 1968.*

Above *This was the first colour transparency I ever took and the notebook reads: 'Transparencies Roll 1. Sunday 6th November 1966. Picture one; 44852 Leeds (Holbeck) shed 55A'.*

Below *. . . and this was my second colour transparency taken minutes later of 'Jubilee' No 45593, Kolaphur, with a 'Black 5' in the background.*

The last five days of our September 1967 tour were spent on the legendary Shap Bank. We stayed with the Thackrays at High Scales Farm, which overlooked the line. There was quite a lot of steam over the bank, especially at night, and it was wonderful to lie in bed and hear the approaching freight's whistle for a banker in distant Tebay and follow the laboured progress of the two engines as they fought over the five-mile drag. Here, Stanier '8F' No 48735 climbs northwards with a freight, banked in the rear by a BR 'Standard 4'. We spent a lot of time around Scout Green signal box, made famous by Eric Treacy's exploits on Shap during the 1940s.

Shap and Beattock

Stanier 'Black 5', No 45013 storms the legendary Beattock Bank between Carlisle and Glasgow with a mixed freight, banked in the rear by a Fairburn 2-6-4T, one of six allocated to Beattock for banking duties.

'Jubilees': 1

Above and left *It was a Sunday in May 1965; we had been stocktaking at the warehouse, and on the way home called in to the Midland sheds. There was a 'Jubilee' in the yard, No 45573, Newfoundland. By this time, 'Jubilees' were becoming rare on the Midland main line. It was an engine with which I had grown up. She had belonged to Leeds Holbeck for as long as I could remember and had been a frequent performer on the London, Leeds to Bradford expresses during the '50s. I clambered all over the engine, probably knowing this would be the last time I'd see her. She was withdrawn later that year, and broken up by Cashmores of Great Bridge in January 1966.*

Below *My original notes on the back of this print read: 'Nameplate of "Galatea", No 45699, in Derby Works after its crash near Bristol.'*

I seem to remember that Galatea was involved in an accident, but don't remember 'Jubilees' ever being overhauled at Derby. The memory fails; but presumably my notes are correct.

Right *No 45665,* Lord Rutherford of Nelson, *a true child-hood favourite, and one of the pictures made during that first afternoon's photography described on page 27. A Kentish Town (14B) engine, she was one of several 'common' 'Jubilees' transferred to Scotland and in return we received some of the 'hardest' Scottish examples. For hundreds of us it was a monumental occasion, and I spent days at Newton Harcourt picking up the 'rarities' as they duly appeared.* Lord Rutherford of Nelson *went to Glasgow, Corkerhill and I don't remember ever seeing her again. She remained there until withdrawal in October 1962 and was finally broken up by Campbells at Shieldhall in December 1963.*

Below *No 45562,* Alberta, *was a Leeds Holbeck engine when I began spotting in 1949, and she remained there until her withdrawal in October 1967. A picture made at Stockport Edgeley shed.*

Below *In the very finest of company: my favourite type of locomotive along with my lifelong spotting friend, George Brunavs, with whom I shared so many epic adventures in pursuit of an all-absorbing pastime.*

'Jubilees': 2

What is your favourite class of engine? Every week someone asks and the answer is always the same; the 'Jubilees'. I was captivated by their shapely appearance; spinning 6ft 9in diameter wheels; three-cylinder rhythms and intriguing names commemorating great achievements in British history — battles, generals, warships and our colonial possessions around the world. Never will I forget their lovely three-cylinder beat as they tackled the climb southwards through Newton Harcourt. On damp days the sound acquired a beautiful clarity which could be likened more to a musical experience than the sound of a locomotive. They were allocated throughout the LMS network from Perth to Bristol and many had an elusiveness which was almost mystical — exactly as did the related un-rebuilt 'Patriots'.

The 'Jubilees' were distributed between 23 depots, many of whose workings were highly specialized, and unless one visited the far away routes over which the engines worked — and spent days by the lineside — there seemed little hope of every seeing many members of the class.

Yet the reality was more subtle; the 'Jubilees' were great roamers and not infrequently drifted unashamedly off their regular haunts, either as a result of depots 'borrowing' them, special workings — both passenger and freight — or their particularly delightful habit of turning up on a rare working, in place of

'Silver Jubilee' No 45552, one of the 'rarest and most atmospheric' of the Jubilees. She emerged from Crewe in 1935 in experimental shining black livery with chromium-plated boiler bands, dome and metal numerals. Red and subsequently green liveries were applied but No 45552 retained her chromium cab-side numbers until she was withdrawn in October 1964.

the usual 'Black 5' or 'Crab'.

How we, in the Midlands, revered rare Central Division ones; *Glorious, Prince Rupert, Conqueror*, or Scottish ones; *Cyclops, Agamemnon*. Yet conversely, northern enthusiasts were enraptured by visions of Kentish town rarities; *Malta G.C., Hawkins, Tyrwhitt* — engines which were part of our daily lives.

The magic of the 'Jubilees' was collective; 191 of them spread across the vastness of our railway network as it was in the post-war years, with all its multiplicity of varied workings and diagrams. They were beasts of the wild to be hunted, scented out and dreamed of; only then did their elusive magic shine.

The few 'Jubilees' preserved today as tourist revivals bear no hint of the former allure of the class: in truth they are extinct, except in the treasured memories of those privileged to have known them in their collective glory.

Not a 'Jubilee' of course, but rebuilt Royal Scot 4-6-0, No 46128, The Lovat Scouts. *It was to this basic form that a few 'Jubilees' were converted, utterly transforming their traditional beauty. Had steam continued, possibly more would have been done but fortunately this didn't happen.*

Royal Scots

Above *Who could forget the thrill of visiting Crewe works in the 1950s? The vastness of the shops with seemingly hundreds of engines embracing dozens of different classes. Fortunately my parent's old box camera went on one of those trips to record this scene of rebuilt Scot No 46126,* Royal Army Service Corps, *in the process of shopping.*

Below and above right *We couldn't believe that all the Scots had gone, not after their performances over the West Coast main line — some of which were almost as scintillating as the present day West Coast electrics. But, by summer 1965, only two remained: No 46115,* Scots Guardsman, *and No 46140,* The King's Royal Rifle Corps; *both at Carlisle Kingmoor. I made a pilgrimage to see them and here No 46140 is depicted in Kingmoore Yard having come in on a pick-up freight from Glasgow (Moss End).*
No 46140 was withdrawn in November of that year and broken up by McWilliams of Shettleston in March 1966.

Below *Happier times with the Scots are recalled by this scene of No 46152,* The King's Dragoon Guardsman, *beneath the electrification catenary at Crewe in 1964, when the engine was allocated to Holyhead. Interestingly, she was the original 'Royal Scot', No 6100, as the identity of the two engines was exchanged in 1933 for 'Royal Scot's visit to America, and the engines never reverted to their former numbers.*

The 'Crabs'

'All the Crabs!' read the piece of paper handed to me following one of my audiovisual shows in Preston in 1986. I had said during the show that I didn't know anyone who saw all 245 LMS 'Crabs', and the gentleman proudly presented himself along with his name and address written beneath the slogan.

It was always a thrill to be at the lineside and see an approaching 'Crab'; all were potential rarities quite apart from their period styling being so pleasing. Though allocated to forty depots, from Scotland to the south of England, the 'Crabs' frequently ran astray; they were the engines of football specials, holiday excursions, summer extras, along with a whole range of freight duties, including pigeon and broccoli specials. Such operations often put them miles off

their beaten tracks, and they were sufficiently utilitarian to be borrowed by recipient depots — sometimes for days on end — to appear in unlikely haunts to the joy numerous spotters.

Typical of countless incidents was one morning during break at school: it was 10.45 a.m. and I had been to the tuck shop when the cry went up, 'There's a Fleetwood "Crab" in the shed yard!' The number was looked up by many occupants of the playground and in the instant, a group of us grabbed our bikes and, oblivious of the consequences pedalled pell mell down the London Road and up to the Bird Cage with its commanding view of the depot. Others possessed of less temerity followed at lunchtime. Such was the magnetic pull of the 'Crab'.

Left *No 42710, one of Newton Heath's rare birds caught at Stockport in 1965.*

Below left *A picture which epitomizes the period styling of the Crabs — a tantalizing mixture of Land Y and Fowler aesthetics.*

Below *A typical 'Crab' scene complete with special plate on smokebox door.*

Stanier 'Crabs'

The related 'Stanier Crabs' had much the same lure as the 'Crabs' proper. Although there were only forty of them, they were similarly elusive. In 1950, only eight depots had them; Nuneaton, Aston, Crewe South, Speke Junction, Brunswick, Mold Junction, Birkenhead and Longsight.

They were seldom seen by day, often being engaged on night-time freights, especially, I suspect, spasmodic traffic related to docks. As with the 'Crabs' proper, they were always taken seriously, but the added intrigue was that from a distance, one could resemble a Jubilee.

Many a time we were spotting from the north end of Rugby Station and a 'Stanier Crab' would come into view with a mixed freight and be held by signals, sometimes for an hour or more. Everyone would mistake it for a 'Jubilee' which, on freight, could mean anything. Unable to restrain our passions someone would volunteer to leave the station and cycle up past the British Thompson Houston Works to find out what it was; his return half an hour later with the news that it was a Crewe 'Stanier Crab' would be the anticlimax of the day. Their elusiveness is summed up by the fact that I saw only 38 of them; failing with No 42948 and 42969 — Brunswick and Birkenhead engines respectively.

Above left *The 'Crabs' proper were built between 1926 and 1932 under Fowler, although their design came from Hughes of the L&Y. When Stanier took office in 1932, he produced his own version, of identical power output, popularly known as 'Stanier Crabs'.*

Above *The similarity of the 'Stanier Crab' design to that of the 'Jubilee' is indicated in this study, and one might reasonably guess that their chimneys were identical.*

Latter-day moguls

Once World War 2 was over, H.G. Ivatt, the new chief mechanical engineer of the LMS, introduced these two classes of mixed traffic moguls to replace a plethora of small and ageing pre-grouping designs. The new moguls broke with tradition by including a number of American characteristics including high foot-plating for ease of maintenance along with many labour-saving devices.

Building of both classes continued until as late as 1952 — four years after Nationalization — by which time the '6400's were perpetuated as the almost identical BR Standard '78000's, whilst the '3000's precipitated the very similar BR Standard '76000's of identical power.

Left and below left *Ivatt '6400' Class 2-6-0 No 46434 shunts the yard at Carlisle Upperby whilst sister engine No 46513 acts as station pilot at Carlisle Citadel.*

Below *Ivatt '3000' Class No 43121 approaches Carlisle from the north with a pick-up freight.*

The Fowler look

Fowler parallel '2300' Class 2-6-4T, No 42374, on the coaling road at Stockport Edgeley depot on 10 April 1965.

What fine period pieces Fowler's 'Parallels' were. We had two at Leicester, No 42330/1 and these were often rostered on the Leicester to Rugby service. I travelled this route regularly on visits to the West Coast Main Line, and remember the magnificent turn of speed the 'Parallels' achieved between stations, not least on the long romp from Ullesthorpe into Rugby. During my travels I managed to see all but one. The exception was No 42415, for years a Greenock engine, which was later transferred to Bangor — an almost equally hard location for anyone living in the Midlands. The 'Parallels' were a key design in the evolution of the suburban/cross country mixed-traffic tank and 125 of them were built between 1927/32.

We called Fowler's seventy smaller 2-6-2Ts the 'Early Tanks', as they began the LMS's number list. LMS No 1 becoming 40001 after Nationalization. But as spotters we always disregarded the four and also the noughts, referring to them as 'Tanker 1' to 'Tanker 70', etc. They were much easier to spot than the 'Parallels' as many were based in the London area for operating suburban trains and performing empty stock movements. I cleared them easily, as did most of my friends, and it was a common sight at Leicester to see grimy examples from the London area, running light engine bound for Derby Works, only to return spotless and emanating lovely wafts of fresh paint some seven weeks later.

Brian and I were on one of our regular visits to London sheds; it was Sunday 29 September 1957 and that day we did Willesden, Old Oak Common, Cricklewood, Stratford and Plaistow.

Here — thirty years later — is Brian's account of an event that day. 'At Willesden you were looking for "Tanker 64" — the last of your "early tanks"; it wasn't on the shed so you suggested going to the foreman's office to ask where the engine was. The foreman was on the telephone reading in detail the whereabouts and status of Willesden's allocation. We stood and waited. You lit a cigarette. Ten minutes later the foreman put down the phone and asked our business. "Where is No 40064?" you said. The reply was terse; "Shunting the carriage depot north of here on the opposite side of the main line."

'It took us a long time to find the place but eventually we scrambled down a bank into a huge yard and there she was, 40064 your last "early tank" and a cop for me too. And I remember you took a photograph in celebration...'

The Fairburn look

Compared with the Fowler 'Parallels', these Fairburns were much less sought after. Aesthetics played an enormous part in the trainspotting process: never was it just collecting numbers; the sophisticated approach taken by most serious enthusiasts lay in the appreciation of the family history of the class; its design characteristics; distribution and nature of work — in that order.

The Fairburns were an updated version of the 'Parallels' but infinitely more modern in appearance. All were built between 1945 and 1951 and accordingly they were too new for true appreciation, their form not having really matured into any established lineage.

Their distribution, however, posed a superb challenge as, apart from being widespread throughout the LMS network in England and Scotland, the type was extensively allocated to the Southern, with a few engines on the North Eastern region as well. It was these latter-day LMS designs which set the precedent for the national distribution of mixed traffic types, as brought to ultimate fruition by the oncoming BR Standards.

As Ivatt's Moguls were perpetuated by BR as standard designs, the Fairburns became the blueprint for BR's standard '80000' 2-6-4Ts, and within months of Brighton Works building the last Fairburn in 1951, they commenced the first batch of the BR version.

Left *The pleasing symmetry of the Fairburn 2-6-4T is indicated in this study of No 42084 — a Darlington (51A) engine.*

Below left *No 42128 was one of the extensive batch allocated to the Scottish region. She is seen here being coaled at Edinburgh St Margaret's depot.*

Below *No 42055 was also a Scottish engine but was re-allocated to Bradford Low Moor during her final years.*

'IF's and 'OF's

Left *Former MR 'early' 'Jinty' 0-6-0T, No 47227, was the only example of this class allocated to Leicester. She was transferred from Cricklewood and still carried her condensing pipes which had been used for working tunnels beneath London.*

Middle left *The old 'IF' tanks were great favourites, especially those which retained their round top Johnson boiler and Salter safety valves. They were the predecessors of the famous 'Jinties', and 240 examples were built by Johnson for the Midland Railway between 1878/99.*

Many survived well into the '50s and I remember them being outshopped from Derby Works. Here, two of Derby's allocation of eight engines — Nos 41747 and 41773 — repose in the shed yard.

Below *Amazingly, five 'IF' tanks survived on shunting duties at Staveley Ironworks until 1966, one of which was No 41763, caught in her final months of service following a special visit I made to see them.*

Right *Another Staveley 'IF' tank was 41712, which came south for breaking up at Cohen's scrapyard at Kettering in 1965. She retained her original Midland Railway chimney and Johnson half-sectioned cab and is seen here in the company of a condemned LMS '4F' 0-6-0.*

Below *'When I finally get ''0F'' tank No 41516 from Burton', my friend Brian Stafford said, 'I will take you out to dinner to celebrate'. His engine was one of three survivors from a class of 28 diminutive 0-4-0 saddle tanks, which the Midland Railway used for shunting in docks and brewery yards etc. Most were scrapped before 1930 but three survived in 1950: Nos 41516/23 were retained for shunting the brewery yards at Burton-upon-trent, but were extremely hard to locate, whilst No 41518 was at Hasland.*

What No 41518 was doing in Leicester when I took this picture is a complete mystery, as normally the engine never left the Chesterfield area. She obviously wasn't on her way to Derby Works so I can only guess that she was en route to a private scrapyard — a practice which was not common at that time.

Brian eventually saw No 41516 and I got my dinner — a fish and chip supper in South Wigston! Nos 41518/23 went in 1955, whilst No 41516 lingered until 1958.

'Black 5's: 2

Above *No 45225 bearing 'Stockport Edgeley' on her buffer beam — as well as a 9B shed plate — receives attention on the ashpits at Leicester Midland.*

Above right *The splendid lines of a 'Black 5' are well-revealed in this study of No 45442 in the shadow of Carlisle Kingmoor's imposing shed clock.*

Right *No 44883 was a Carlisle Kingmoor engine for as long as I can remember, and one which occasionally turned up at Leicester with the daily Kingmoor working.*

4-4-0s

One of my happiest memories was sitting on the bridge at Newton Harcourt, listening to the musical three-cylinder throb of a compound tackling the climb southwards from Kilby Bridge. Invariably, one would pass after a southbound express — the compound following with an all-stopping train to Bedford, or even London St Pancras. On clear, damp evenings, the effect would be ecstatic and, as the three-cylinder rhythms of the 'Jubilees' receded to the south, the muffled throb of a compound became audible from the north. For a few blissful moments, the compound and 'Jubilee's rhythms would be mixed in a tantalizingly beautiful interplay, until the latter subtly gave way.

During the very early years of the '50s, some of the compounds were original Midland ones, Nos 41006/11/41 being Leicester engines, whilst No 41000 herself was a Derby engine and regularly worked south — occasionally as pilot to a 'Black 5' or 'Jubilee' — on a Manchester Central to St Pancras express.

It was a very different experience to witness the Caledonian Railway 4-4-0s and although their lineage was a noble one, having been descended from the famous Dunalastairs, their duties over the final years seemed somewhat less flamboyant than their Midland counterparts, although I did often see them piloting 'Black 5's on the Perth to Inverness route. There were 48 of them, and I managed to see all but four — a fine achievement considering they never worked south of the border and were split between fourteen different sheds.

Below left *Former LMS compound 4-4-0 No 40934, a Gloucester engine, awaits its return working at Leicester Midland. A picture made during the months I worked at Leicester Midland sheds in 1957.*

Below *During our Scottish trips we saw these lovely 4-4-0s, introduced by Pickersgill for the Caledonian Railway between 1916 and 1922, as direct descendants of the earlier Dunalastairs. A scene at Forres in 1956.*

'2F's and '3F's

Above *Ex-Midland Railway '2F' No 58298 was a long-time Leicester favourite and still sported her original Johnson cab.*

Left *No 43326 was one of fourteen former Midland Railway '3F' 0-6-0s allocated to Leicester.*

To have grown up amongst these former Midland Railway inside cylinder 0-6-0s in line service was something of a privilege. A Leicester '2F' had a daily working south along the Midland main line whilst '3F's were regularly employed on lighter freights, often over long distances. Many however were relegated to shunting and our examples could be found at Leicester North sidings, Knighton and Wigston. My favourite Leicester '2F' was No 58164 which came to us from Kettering during the early '50s. She had massive reddish-cream numbers on her cabside and somehow looked much older than her sisters. I used to love seeing that engine. I remember once going to a Leicester clinic to have a particularly nasty verruca removed; I was terrified, as the pain was acute, so I determined that during the operation, I would dream about No 58164. I went to sleep under the gas with images of that lovely engine residing amid the sooty silence of Wigston Roundhouse on a Sunday afternoon, where I used to go and clamber all over her.

Above *No 58305 was a Leicester '2F' which came from Bedford during the mid-'50s.*

Below *No 43728 — Another Leicester '3F' — stands resplendently in Leicester Midland yard before working a special train over the former Leicester and Swannington line during the summer of 1956.*

'4F's and 'Wheezers'

The archaic North-Western front of the 'Wheezer' 0-8-0 was unmistakable, even from long distances.

A brace of the Midland Railway's classic '4F' 0-6-0s, no less than 772 of which were built between their inception in 1911 and 1940.

No one who remembers hearing an LNW 'Wheezer' in full cry will ever forget the sound; two long exhaust beats followed by two short ones, accompanied by the most hideous wheezing from the cylinder glands — the sound resembling the crying of a baby. You could hear the 'Wheezers' coming a mile off, their stark bare fronts invariably shrouded in leaking steam. They were terrifying and upon hearing one climbing the grade from Kilby to Newton, I used to run behind the fence and hide until the engine had passed. Their appearance was archaic; they were uncomfortable to work on and prone to slipping, yet, for sheer slogging hauls, they had few equals.

The '4F's are another class popularly disparaged today; they were out of date when they were introduced, critics will say. But after fifteen years of lineside siding, I developed a great respect for them. They could be found from the Highlands — where they had lovely capouchon chimneys — right down to the Somerset & Dorset system where they frequently worked passenger trains and even piloted Southern Pacifics over the tortuous Mendips. The '4F's came from a time when our railways were properly utilized and there was enormous demand for medium-sized inside cylinder 0-6-0s which could work almost anywhere. Excellent on shunting, tripping, intermediately-sized main line freights and excursion passenger trains alike, the '4F's were cheap, reliable and efficient, and played their integral role with distinction.

As an example of their prowess, I recall an occasion in 1963 when a diesel on the Waverley express failed, and '4F' No 44386 took the 300-ton train forward from Appleby to Carlisle 30¾ miles in 34½ minutes, start to stop, with a top speed of 80 mph near New Biggin.

Consols '8F'

Trainspotters in the Midlands were particularly well-blessed for seeing Stanier '8F's — or 'Consols' as we called them — although I only know one person who cleared them. I finished needing twelve, most of which belonged to south Lancashire and Cheshire sheds, intensely industrialized areas where workings were relatively localized. I also needed two from Shrewsbury and one from Swansea Victoria. The most amazing deficiency however, was No 48209, a Canklow (19c) engine, a shed which in all my childhood endeavours was never visited. Yet, considering the years I spent in the Midlands area, it is amazing how this '8F' eluded me. Certainly Canklow's '8F's were quite rare through Leicester, but they did occur and

my failure to run this engine to ground over so many years is remarkable.

I have always held the '8F' in special regard, ever since that dramatic experience at Newton Harcourt in 1949. They were the commonest type on the Midland main line — Toton shed alone having fifty of them. They were employed on all kinds of freight work but especially on the famous Toton to Brent coal hauls and ironstone workings northwards from Northamptonshire — duties which they shared with the LMS 'Garratts'.

Introduced in 1935, the '8F's had boilers and cylinders identical to those of the Stanier 'Black 5's, and their ranks were swelled in 1939 when they were

chosen by the Ministry of Supply for war needs. This led to the '8F's being built by other workshops; Eastleigh, Brighton and Ashford on the southern, Darlington and Doncaster on the LNER and Swindon on the Great Western, in addition to Crewe and Horwich on the LMS. Others came from private builders. After the war, 663 came into British Railways service and, like the 'Black 5', became spread over most of the LMS system. They were amongst the last steam locomotives to be withdrawn by BR as, like the 'Black 5's, they had years of useful life left in them and the sight of the once prestigious '8F's down-graded to standby duties was utterly incongruous to all who knew them in their prime.

Below left *Stanier '8F' No 48728 was a Leicester engine through much of her working life. She's seen here on shed at Burton-upon-Trent.*

Below *Stanier '8F' No 48606 also residing outside Burton Shed.*

The Great Central's last days

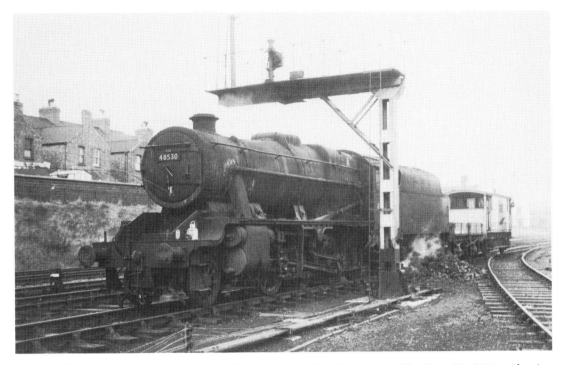

Above *By 1965 the Robinson/Thompson 2-8-0s had been largely replaced by Stanier '8F's. Here, No 48530 — bearing a 15E (Leicester Central) shed code — with little to do, stands opposite the depot.*

Above left *Stanier 'Black 5' No 45222 waits amid the ornate portals of Leicester Central at the head of a southbound stopping train.*

Below left *Surrounded by classic Great Central infrastructure, 'Black 5' No 44847 leaves Leicester Central with a train for Marylebone.*

Witnessing the deliberate run-down of the Great Central was traumatic. In 1950 it was a thriving system, with a fabulous variety of trains ranging from the luxury Manchester to Marylebone expresses to a wide range of fine cross-country services. Holiday times brought many specials with rare LMS engines from the Central Division; Bulleid Pacifics up from the south, and Halls and Granges from the Great Western. The top expresses were worked by Gresley's 'A3' Pacifics, including *Flying Scotsman*, which was a Leicester Central engine for some years.

Freight traffic was prolific: the constant stream of southbound coal hauls being augmented by a whole range of commodities, not least fish, and it was normal for fish landed at Grimsby in the morning to be frying in Midland's shops the same evening. Proof

enough that motorways and juggernauts are a confidence trick played by those who have vested interests in them.

Of course the Great Central had been infinitely more exciting before I knew it; it was the last main line into London and one of Britain's most immaculate railway companies and when pioneered during the 1890s, was part of Sir Edward Watkins' dream to have through expresses running from the industrial north of England direct to continental cities via running powers over the South Eastern and Chatham Railway and through the proposed Channel tunnel, on which work had actually started!

The Great Central closed in September 1966 and the final trains were worked by redundant 'Black 5's, sent to the Central to die.

Chapter 3

Industrials
Doxford crane tanks

During a bitterly cold spell in December 1970, I made an overnight journey to Sunderland to photograph the last crane tanks in British service. They were at Doxford's historic shipyard on the River Wear and were due to be withdrawn the following January. Doxford — celebrated Wearside shipbuilders since the middle of the nineteenth century — reorganized their yard at the turn of the century to a layout based on the use of crane locomotives and the first of these celebrated engines arrived in 1902 from Hawthorn Leslie.

Left Millfield — *one of Doxford shipyard's 0-4-0CTs, built by Robert Stephenson and Hawthorn in 1942 — prepares for the day's work at Pallion Shed, shortly before dawn on a December morning in 1970.*

Right *Sister engine* Roker *at repose in the depot yard during a spell of pale, wintery sunshine.*

Above *With Drakelow power station in the background, Hunslet 'Austerity',* Cadley Hill No 1, *passes* Progress, *an RSH 0-6-0 saddle tank of 1946, in the colliery sidings at Cadley Hill near Burton-upon-Trent.*

Right *Kitson 0-6-0 saddle tank* Caerphilly *along with the redoubtable No 19 on the Storefield ironstone system.*

Below *It was men like Chris Boyle and George Edlin who kept Cadley Hill's steam fleet in such magnificent condition. Cadley Hill No 1 is a 1962 engine and one of the last three 'Austerity' Class locomotives to be built.*

Above *A shale train heads gingerly out of Pennyvenie Mine in Ayrshire, and away through the morning mist behind an Andrew Barclay 0-4-0 saddle tank on 15 May 1972.*

Below *The mists now dispersed, the same Andrew Barclay prepares to take a loaded train out of Pennyvenie Mine, bound for the BR Exchange at Waterside.*

Above *A glorious autumn day spent at Goldington Power Station at Bedford with No ED9, an oil-fired Andrew Barclay 0-4-0 saddle tank — Tuesday, 9 November 1971.*

Below *On Saturday 16 September 1967 — along with Brian Stafford and Judy Maddock — I began a two-week tour of the northern sheds, and during our stay at Lostock Hall on Wednesday 20th we took time out to see the Bagnalls shunting in nearby Preston Docks. Here is* Enterprise, *delivered from the famous Stafford Works to Ribble Navigation in 1948.*

Above *By Friday 22 September, we had visited several colliery networks in the north-east, especially Philadelphia, on the former Lambton network. Sitting amid the mottles of the old shed were a pair of 0-4-0 saddle tanks, a Hawthorn Leslie (left) and a Hudswell Clarke.*

Below *Monday 25 September was one of the happiest days of my life, as we were escorted over the Ashington Colliery network by our dear friend Stoker Redfern, who had retired after a lifetime's service at Ashington. In the works undergoing overhauls was a Hunslet 'Austerity' with Giesl chimney — along with an RSH 0-6-0ST.*

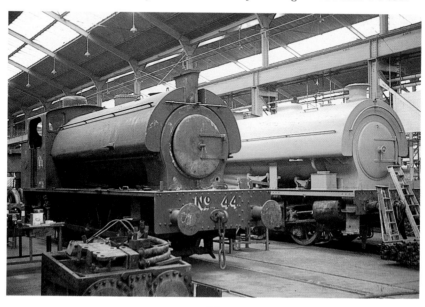

On Saturday 23 September, we were at Backworth, where I made this study of two Hunslet 'Austerities' framed by the shed entrance. No 49 is a Giesl-chimneyed RSH, whilst No 6 is a Bagnall engine sporting the standard chimney.

Castle Donington power station was one of the many industrial locations which preferred to continue with steam, and they kept a pair of these lovely RSH 0-4-0STs in perfect condition; a picture made in November 1971.

Industrials from childhood

Left *My earliest memory of industrial engines, as distinct from those on the main line, was with the examples at Enderby Quarry. I don't remember how I came to find them, but it was probably whilst cycling from Oadby to Nuneaton, for the West Coast main line. I was about eleven, and I remember them as being well-maintained and looking almost new. This picture of* Elizabeth *proves my fondness for them and I seem to think that at least one other had a girl's name — possibly the quarry-owner's daughter, in common with long industrial practice.*

Elizabeth *is the unidentified engine of this book; she is but a shadowy memory of childhood, as are her sisters. If I really wanted to search their origin, I could — doubtless some readers will know it — but perhaps I prefer to keep them shrouded in mystery as a pure nostalgic memory of a set of engines which fascinated a small boy and left an indelible impression.*

Right *But few engines have inspired me more than* Cecil Raikes. *Ever since I heard that this big Mersey Railway 0-6-4T was in colliery service at Heanor in Derbyshire, I nurtured an ambition to go, and, one summer day in 1954, I set off on my bike for the long, hilly journey. I was so inspired that I took my camera for, even at the age of fourteen, I knew the significance of this amazing survivor which once worked through the tunnels under the Mersey between Liverpool and Birkenhead.*

Her time in those smokey tunnels was limited and when the Mersey railway was electrified in 1904, Cecil Raikes *— a mere nineteen years old - was sold to the Coppice Colliery in Derbyshire, where she remained for over half a century. At Heanor I met a kindly man who showed me his written records of* Cecil Raike's *history, and of all the other engines which worked there — if only I had taken a copy of those notes!*

I stayed with the big engine until late in the afternoon before starting my long cycle ride home. It was late and I would be in trouble, as I had an evening paper round to do in a fashionable part of Leicester. It was 8.30 pm when I reached the newsagent's, utterly exhausted; he was furious. His customers had been ringing for their papers and although half had been delivered, I had to go out and finish the rest — and to be chided at every other door.

Northants ironstone 1: Corby and Storefield

Left *One of the giant '56' Class 0-6-0STs specially built for Stewarts and Lloyds Minerals during the early '50s by Robert Stephenson and Hawthorn. They were very successful and hauled trains of several hundred tons from the outlying quarries up to Corby Works.*

Right *No 19 climbs out of the pit and heads for Storefield woods with a loaded train. She had a strange, rasping cough, and was not as strong as her crews would have liked. She was finally broken up by Cohen's of Kettering in 1969.*

Below *Storefield's lovely Andrew Barclay 0-4-0T No 19 built in 1940 waits at the loading siding.*

Corby — new town of the '30s — grew when the steel-works were greatly expanded, drawing workers from all over Britain, many coming on foot to seek employment during the depression. The impetus was the vast reserves of ironstone, first exploited commercially after the building of the Midland Railway main line had revealed the presence of huge quantities of ore.

A vast network of lines spread from Corby Works into the surrounding countryside, but not all the ore was smelted locally; some was conveyed away by rail to the ironworks of North Notts and South Yorks. Store-field was one of these exporting systems: owned by the South Durham Iron and Steel company, the system connected with BR's Kettering to Manton line.

Today, no ironstone mining takes place in Northamptonshire, and Corby Works is reduced to processing imported blooms. The destruction of this once vast industry by the claim that it is cheaper to import ore from overseas is regarded by many who worked in the industry as dubious, if not actually sinister.

Northants ironstone 2: Cranford and Nassington

Left *As afternoon shadows lengthen, and the day's duty is done, Cranford's Avonside 0-6-0ST returns to her depot.*

Right *Foggy day at Cranford. Cranford ironstone invariably evoked memories of sunny days in the lush Northamptonshire countryside, but there were exceptions as this study of their Avonside 0-6-0ST proves.*

Below Jacks Green *and* Ring Haw, *that celebrated pair of Hunslet 16-inch 0-6-0STs, at work in their natural habitat at Nassington Ironstone Mine in 1969. Both engines were named after local woods.*

Some of my happiest days on the Northamptonshire ironstone system were with the late Reverend Teddy Boston. I had just turned professional and Teddy would pick me up from my Leicester bedsitter for days on the ironfield. Cranford was a firm favourite; the system was rustic, friendly and quaint. Wild strawberries grew in profusion by the tracksides, and the industry blended perfectly with the glorious Northamptonshire countryside. The Cranford system connected directly with BR and their ore was taken along the old Kettering to Cambridge line. Inevitably rationalizations swept Cranford into history, but to have experienced it in the company of Teddy Boston is one of my most privileged memories.

Another highspot of those early years was Nassing-ton, which became the last ironstone mine in Northamptonshire to retain steam. I used to hitch-hike along the A47 from Leicester, get down at the Wansford Turn and walk the remaining mile to the mine. Countless happy days were spent there in the company of the engines and men, like Jim Hopkins and Bill Prodger, who had driven these Hunslets for over thirty years.

The people, the engines, the environment, the photography and even the climate (it was usually sunny) made Nassington one of the most wonderful places I have ever known. Today, the site has been reclaimed by agriculture, and the rolling green fields bear no witness to this once superb bastion of industry.

Coal 1: Backworth and Bedlay

Below *A bitterly cold night in December 1971 on the Backworth system north of the Tyne — one of the most exciting colliery networks in the north-east, albeit much reduced in its later years. Backworth was one of the last bastions of British industrial steam and pride of their fleet was this 18-inch 0-6-0ST, one of a generation of 16-inch, 17-inch and 18-inch saddle tanks built by Robert Stephenson and Hawthorns for the north-east coalfield. She was built as recently as 1957 as RSH No 7944; she weighed 53 tons and, by the time this picture was made, was the last survivor of her class. She was more powerful than a Hunslet Austerity and her tractive effort was actually greater than that of an LMS '4F' 0-6-0!*

Right *One of the most distinctive survivors from the latter days of industrial steam was the famous Hudswell/Clarke 0-6-0T at Glenboig Colliery, Bedlay, near Glasgow. Built in 1909 as HC No 895, she was generally similar to the BPGVR engine on page 74, not least in the characteristic chimney design. Again, one can apply the analogy that, if it looks like 'Tank Engine Thomas', it's a Hudswell Clarke!*

I made this picture in May 1972 during a tour of the Scottish coalfield, and after leaving Glenboig, drove south to Backworth, to spend that evening sitting round the eternally roaring fire in the depot mess room listening to epic tales of locomotives and events as inimitably recounted by Backworth veterans. In fact, I slept rough in the workshop that night, in order to be ready for photography by first light.

Coal 2: Hunslet 'Austerities'

Above *Our May 1972 tour of the Scottish coalfield included Polkemmet Colliery in West Lothian which had a fabulous range of motive power, including a Grant Ritchie 0-4-2ST. Double-heading up the bank to the BR connection on Polkemmet Moor was commonplace and the pilot, to our sheer joy, was vintage Andrew Barclay 0-6-0ST No 885 of 1900. She is seen with a Hunslet 'Austerity' purchased by the NCB from the War Department for £1,500. Watching the two attack the climb was enthralling; the 'Austerity' created a steady even blast, whilst the ancient Barclay wheezed and rasped in an entirely different rhythm, so creating marvellous polyphony.*

Left *The essence of a colliery railway in a mining community is captured by this scene of a Hunslet 'Austerity' trundling across the road at Harrington Colliery.*

Right *During the snows of February 1973, I made another pilgrimage to the South Wales valleys and returned to Hafod-rhy-nys Colliery, a favourite haunt. Hunslet 'Austerity' Llewellyn was working the shale bank.*
The derelict structures of a legendary, industrial past dominate the scene and I couldn't help thinking, as I watched Llewellyn playing out her final moments, that the world's first steam locomotive had been born over the hill in the adjacent valley one and three-quarter centuries previously.

Coal 3: The Lambton 0-6-2Ts

Right *A Kitson 0-6-2T passes Shiney Row on the Philadelphia colliery system on Friday 22 September 1967.*

Left and below right *Robert Stephenson 0-6-2T No 42 working at Philadelphia on Friday, 22 September 1967.*

During the '60s, the small mining village of Philadelphia achieved fame as being the last bastion of steam on the once-huge Lambton colliery network which as long ago as 1860 had 70 miles of railway. There were some real gems at Philadelphia including a Lambton-built 0-6-0 tender engine of 1877 — she survived until 1966. However, when my visit was made in September 1967, the most distinctive survivors were the huge 0-6-2Ts from Kitson's and Robert Stephenson's.

In 1929, more 0-6-2Ts were purchased, this time former Taff Vale engines from the Great Western, so giving the system three similar designs of this type One of their regular duties was to work heavy coal trains from Penshaw to Lambton Staithes on the Wear at Sunderland. In 1967, a pair of 0-6-2Ts received heavy overhauls at Lambton Works: this raised hopes for the future but, by 1969, all these steady and hardworking locomotives had been withdrawn.

Chapter 4

British Railways

The Pacifics

Left *Visits to Carlisle in 1965 revealed the last 'Clans', a class of ten lean Pacifics built for the Scottish region in 1952 as a lighter version of the 'Britannias'. They were not particularly successful, being heavy on coal, and although more were planned — including some for the Southern — they were never built.*

No 72007, Clan Mackintosh, was one of the last to go. She remained at Carlisle Kingmoor through that year but was broken up by Campbell's of Airdrie in March 1966.

Below *Throughout most of their working life, the first five 'Clans' were allocated to Glasgow Polmadie, and the second five to Carlisle Kingmoor. Polmadie's five were withdrawn en-bloc in October 1962, after a working life of only ten years. All were stored at Parkhead depot before being taken to Darlington Works for breaking up during the early months of 1964.*

Kingmoor's engines fared better and here is No 72008, Clan Macleod, on shed in 1965. She was withdrawn in April 1966, and finally disappeared into McWilliam's Shettleston Yard the following June.

Above *The 'Britannias' distinguished themselves throughout Britain over a short but scintillating career, during which they appeared on all regions. Perhaps their most celebrated work was in East Anglia, especially with the two-hour expresses between London Liverpool Street and Norwich. No 70005,* John Milton, *was one of those engines, formerly allocated to Stratford in East London. In the later years she joined many of her sisters at Carlisle Kingmoor where, by the summer of 1967, an amazing 41 Britannias — out of the 55 built — were allocated.*

Below *We always thought of* Duke of Gloucester *as a replacement for LM Pacific 46202,* Princess Anne, *which perished in the Harrow and Wealdstone disaster of 1952.* Princess Anne *was an accountant's rebuild/replacement of the earlier* Turbomotive, *which was effectively withdrawn in 1950.*

Less than two years after Princess Anne *disappeared into Crewe Works,* Duke of Gloucester *emerged, complete with three cylinders and Caprotti valve gear — the last Pacific built for British Railways. The engine was allocated to Crewe for express service on the West Coast main line, and showed great promise. But by this time, railway management was so preoccupied with impending dieselization and electrification that the* Duke *exerted little influence and was withdrawn prematurely in November 1962, having been in service for a ridiculously short eight years.*

'9F's

Left *A misty morning scene between Great Glen and Newton Harcourt on the Midland main line, as a BR '9F' 2-10-0 romps northwards with a freight.*

Below *When BR introduced the first '9F's for heavy freight working in 1954, they were greeted with much opposition. There was particular resistance on the Great Western, where some drivers refused to work them, preferring to continue with their '2800's.*

Many enthusiasts disliked them too, and I remember the disdain amongst Leicester spotters, who regarded the '9F's as tinny, and thought the Stanier '8F' superior in every way. Here is No 92139, built at Crewe in 1957 in Leicester shed yard with Hillcrest, a former Victorian workhouse, in the background.

Whatever disdain the '9F's encountered initially was rapidly disproved. Quite apart from being powerful, free-steaming engines for heavy freight work, they proved capable of sustained fast running, too. So capable in fact, that for a while Leicester used them on express passenger workings to St Pancras, which involved average speeds of 60 mph, including sprints into the 90s, until the practice was forbidden by Derby HQ.

Paradoxically, although the '9F's were the only non-mixed traffic design produced in BR's twelve Standards, they proved to be the most versatile of them all. Certainly, no steam design in British history had been so capable of performing such highly contrasting duties. This scene shows No 92018 being turned in Leicester roundhouse, and subsequently leaving the shed in readiness for taking a freight to her former home depot of Wellingborough.

Mixed traffic: 1

Right *The last of the archaic Midland Railway '2F' 0-6-0s survived at Coalville until 1964, by virtue of their being needed for working coal trains over the old Leicester and Swannington route which included the notoriously narrow bore of Glenfield tunnel — the '2F's being the only type capable of negotiating it. But the sheer difficulty of maintaining the '2F's forced action to be taken, and two of BR's Standard '78000's were given specially reduced cabs to perform the service and allow the '2F's to be retired. The engines were Nos 78028/9 and here the former is depicted shunting at Desford during the spring of 1966.*

Below right *The smallest Standard design was the '84000' Class 2-6-2Ts, based on Ivatt's earlier '1200's for the LMS. Many were motor-fitted for push and pull workings on branch lines, but only thirty were built due to the rapid encroachment of diesel railcars combined with the practice of closing the very lines for which the type was intended.*

Below *Bournemouth shed in 1965 with a BR Standard Class '4' in quiet repose. Visibly the descendants of the earlier Stanier and Fairburn 2-6-4Ts, they were allocated throughout Britain. The Southern examples remained on the Waterloo, Bournemouth and Weymouth section until July 1967, having long since replaced older Southern classes on semi-fast and stopping trains.*

Mixed traffic: 2

Above *I spent much of the final weeks of British steam in Lancashire, visiting the last shrines; Patricroft, Rose Grove, Lostock Hall and Carnforth.*
It was traumatic. Everything we had dreaded for so long was about to happen and cameras clicked constantly in a vain bid to compensate for the treasures which had been lost over those long golden years when steam was going to last forever. There were many special rail tours, and one was double-headed by Standard '4's No 75019/27. The pair are seen approaching Skipton on Sunday 28 July 1968.

Below *Another picture made under deeply sad circumstances on the final day of Southern steam, on Sunday 9 July 1967.*
We were on Wishing Well Bank above Weymouth, to see Merchant Navy No 35030, Elder Dempster Lines, on the 14.07 express to London Waterloo, which we thought would be the last steam diagram out of Weymouth. But we remained on the bank that afternoon, miserably watching the light engines drifting down to Weymouth for disposal when, to our sheer joy a steam train was heard climbing the bank. It was BR Standard '5' No 73092, at the head of a Channel Island's fruit special from Weymouth docks to Westbury — my last picture of active steam on the Southern.

During those final days of steam, the subtle distinction between the middle range of BR Standard designs were ignored, the best survivors of the various classes being randomly transferred to wherever they were needed. On the Southern, we noticed that the Standard '4' 4-6-0s, Standard '3' 2-6-0s and Standard 2-6-4Ts were interchangeable. Here, Standard '3' No 76006 heads a stopping train west of Wareham on 27 March 1967.

Condemned

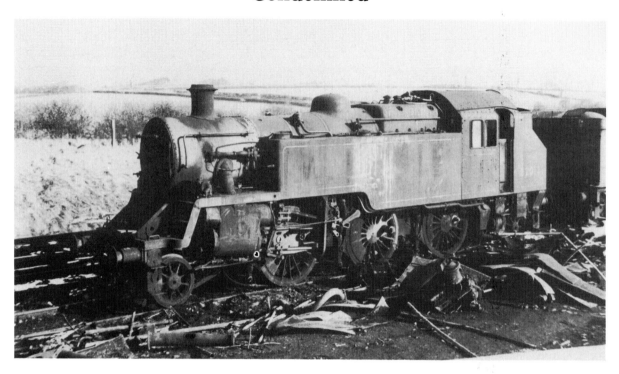

Following the nationalization of Britain's railways in 1948 the engineer Robert Riddles supervised the production of a range of twelve standard steam designs intended to replace the innumerable types inherited from the former companies and take Britain's railways forward to the twenty-first century. No sooner had they been put into traffic than the decision came to abandon steam and Riddles' designs went to the scrapyard with the older classes they were meant to replace.

It makes me feel very old to think that I was watching trains years before the Standards were designed; I saw many of them built, witnessed their decline and yet work today alongside members of a generation born in the post-steam age who regard the steam locomotive — if they regard it at all — as something belonging to the distant past.

Of course the true perspective is that I am not old; the acute changes inflicted upon Britain's railway industry over the past thirty years would have been utterly and totally unimaginable as recently as 1950. Few of the changes wrought were for the better, but the most bizarre of all was the sublimation of our railway infrastructure, traditions and heritage, in favour of the insanity of basing the national economy upon road transportation; action which will surely be seen by future generations as one of the twentieth century's most destructive acts of folly.

But looking at the picture of No 73013 opposite — taken over twenty years ago — I think of the time when I saw that engine being built at Derby. Regular monthly visits to Derby Works during the early '50s ensured that we not only saw engines from all over the country receiving overhauls, but also caught the batches of new Standard '5's as they appeared between 1951 and 1954.

No, I am not old at all; the Standards came and went in a few short years, despatched in the mania for

96

Above left *BR Standard '2' 2-6-2T No 82012 awaits breaking up at Cohen's scrapyard in Kettering.*

Above right *For many years, BR Standard Class '5' No 73013 was a Shrewsbury engine, and after a spell at Willesden, returned to GW territory in 1965 when she came to Banbury, but was eventually put in store.*
In May 1966, she was transferred to Bolton but did little, if any, work there, as the engine was broken up by Cashmore's at Great Bridge in August.

mass dieselization at whatever cost. And, with the trusty steam locomotive — with all its inbuilt simplicity and longevity — went half of Britain's railway network too: an action which quite cold-bloodedly left the surviving half impotent to fulfill the nation's true transport needs.

Chapter 5

Southern
Southern Pacifics

Rebuilt Bulleid 'West Country' No 34013, Okehampton, *was one of the engines which survived until July 1967.*

It was remarkable that the smallest of the big four companies, and the first to embark upon extensive electrification—some as early as the 1920s—should be the last to retain steam on fast long distance expresses. But, until that fateful day of 9 July 1967, virtually all trains from London, Waterloo to Southampton, Bournemouth and Weymouth remained steam-hauled, primarily by Bulleid's magnificent Pacifics.

Quirk of circumstances allowed Bulleid to produce his revolutionary 'Merchant Navy' Class in 1941, when the country was at war, and these were followed in 1945/6 by the lighter 'West Country' and 'Battle of Britain' classes, building of which continued until 1950. Superb as both classes were, a decision was made to rebuild them in 1957 and, when the work finished in 1961, sixty light Pacifics had been converted, which left fifty running in their original condition. All the 'Merchant Navy's were converted *en masse* in 1959.

This rebuilding was undertaken at a crucial point in British motive power policy; some Standard designs were still being constructed, and dieselization had not gained sufficient foothold. Thus, at the eleventh hour, the Southern had equipped itself with a modern stud of steam locomotives which rightly remained in full command until finally ousted by the Waterloo to Bournemouth electrification.

Another rebuilt 'West Country' No 34108, Wincanton, undergoing overhaul at Eastleigh Works. Wincanton was a regular Salisbury engine until withdrawal in June 1967. She was despatched to South Wales along with many of the Southern's engines later that year and broken up by Buttigieg's of Newport in 1968.

Waterloo to Weymouth

Above *The grassy banks at Worting Junction, to the west of Basingstoke, were one of our favourite places for enjoying Southern steam. It was at Worting that the West of England line for Salisbury and Exeter diverged from the Southampton, Bournemouth, Weymouth route and, by positioning ourselves east of the junction, we saw a tremendous variety of trains, especially on summer Saturdays, when many specials would be running. This 1964 picture captures some of the atmopshere, as Brian Stafford watches rebuilt 'Battle of Britain' No 34088,* 213 Squadron, *pass with a special boat train from London (Waterloo) to Southampton.*

Below *Waterloo Station in 1965; the last of the great London terminii to retain steam-hauled expresses. With soft, rasping exhaust beats, so characteristic of the 'Merchant Navy' class, No 35003,* Royal Mail, *pulls away. The white discs on the engine's front indicate the train to be a Waterloo to Bournemouth express.*

We spent countless happy weekends at Poole with Judy Maddock's family over the final years of Southern steam, and often ventured westwards through the quiet Dorset countryside towards Weymouth. One such occasion was on Monday 27 March 1967, when, during a spell at Wareham Station, I made this study of 'Merchant Navy' No 35030, Elder Dempster Lines, with an up express. The driver is not wearing glasses but goggles to protect his eyes from smuts!

Maids of all work

Left *There was a fair amount of freight on the Southern's West of England main line during steam days, and if we were lucky, one or two of these handsome S15 4-6-0s could be seen during a day's linesiding.*

Introduced by Urie for express freight work on the London and South Western Railway in 1920, the type was continued under Maunsell, and No 30837 belongs to this later batch dated between 1927/36. The S15s were excitingly reminiscent of the 'King Arthurs', so providing us with a glimpse of period Southern design.

Left *Southern Railway 'N' Class 2-6-0, No 31842 is of particular interest in being one of fifty engines of this class built at the Woolwich Arsenal as part of a government measure to alleviate unemployment there towards the end of World War One.*

The Woolwich engines were later taken into Southern Railway stock, but, due to the inexperience of the Arsenal's men in building locomotives their engines were not the equal of the Ashford-built examples until they had received their first major overhauls.

Above right and right *Some names from withdrawn 'King Arthur' Class engines were given to the Southern's BR Standard '5' 4-6-0s between 1959/61. Twenty were so named, including No 73080, Merlin, and No 73115, King Pellinore. There was a popular belief that these were the original 'King Arthur' plates; the shape and lettering was similar, but the 'Arthur' plate carried the name of the class as part of the casting. Even so, the Standard '5' plates were highly prized and were removed before the engine's withdrawal to prevent theft — note the right-hand bolt of* King Pellinore *plate is predictably missing!*

Southern tanks (indigenous)

Above *We had got into Eastleigh Works under a broken wire fence; it was a quiet Saturday afternoon and few people were on duty. We covered the works yard without being seen before entering the erecting shop. There was always a lovely smell about erecting shops, especially when they were still, the only sound being a hiss of compressed air. Many Bulleid Pacifics were in various stages of overhaul, along with several older Southern classes, but the pride of the shop was one of Dugald Drummond's lovely M7 0-4-4 suburban tanks of 1897. She must have been one of the last to go through the works and was presumably a Nine Elms engine used for coaching stock movements around London Waterloo Station.*

Left *The last survivor of Adam's B4 0-4-0 dock tanks, No 30102, formerly named* Granville, *stands redundant at Eastleigh after withdrawal in 1963. Introduced in 1891, the 'B4's were the traditional shunting engine of Southampton Docks, until replaced by the United States Army 0-6-0Ts following the Second World War. Throughout their long history, Southampton's 'B4's were known by their names rather than their numbers.*

Above *Imagine my excitement in visiting Eastleigh Works in 1957 and finding this Plymouth Devonport and South Western Junction Railway 0-6-2 tank shunting the yard. Named* Earl of Mount Edgcumbe, *she was a truly rare find. Sadly, she had come from Plymouth for scrapping and was merely being utilized by Eastleigh on a temporary basis. By the end of that year she and her sister,* Lord St Levan, *had disappeared.*

Below *The Beattie well-tanks were a predecessor of the 'M7's for working suburban trains out of London Waterloo — their introduction in 1874 rendering them one of the first suburban designs. All were scrapped by 1898, except for three engines retained for working China Clay trains over the lightly-laid Wenford branch in Cornwall. The oldest design working on BR, they were trainspotter's gems, and Brian, George and I went to Cornwall to find them in 1960, when this picture was made. They survived a further two years having outlived their sisters by 64 years!*

Southern tanks (American)

'Have you cleared your Southampton dock tanks?' was a regular spotter's question in the '50s. They consisted of fourteen engines allocated to 71I (Southampton Docks) which lost themselves amid the vastness of the dockland area, many parts of which were not easily accessible and others totally forbidden — a frustrating situation as there was always a dock tank hidden behind every corner and few spotters managed to see all of them. They were regarded as a great novelty, being of pure American design and part of the vast postwar dispersal of engines specially built for service during the hostilities.

Almost 400 were built by three American firms — Davenport, Porter, and the Vulcan Ironworks between 1942 and 1944, primarily for service in the European war theatre. Typically American, they had bar frames and first arrived in Britain in July 1942. Some remained in

Britain to work in industries which were expanding through war-time needs, but most went to Europe to work behind Allied lines. After the war, they proved an excellent and virtually new shunting locomotive and dispersal was widespread, with many remaining in the countries they had operated in during the hostilities.

The Southern Railway took fourteen into stock in December 1946, along with an extra one for spare parts, which was officially scrapped in 1950. They were put to work shunting the vast and busy Southampton Dock area, and replaced the ageing LSWR 'B4' 0-4-0 tanks depicted on Page 104. They were considerably larger than the 'B4's and in order to warn crews long accustomed to the 'B4's compact shape a notice was painted on their bunker sides: 'When standing on the footstep you are not within the loading gauge'. They

were a familiar sight at Southampton until gradually displaced by diesels from 1963. Since their demise in Britain, I have found examples in France, Greece, Yugoslavia and China.

Below left *During later years, when diesels encroached into Southampton Docks, some of the former USATC 0-6-0Ts moved to other parts of the Southern system and No 30072 became Guildford Shed's pilot, remaining on this duty until July 1967.*

Below *Other examples acted as works pilot at Eastleigh as, by 1964, all the older London and South Western Railway types suitable for this function had been withdrawn.*

Isle of Wight

The Isle of Wight's glorious remove from the hurly burly of life on the mainland was doubtless one reason for the survival until 1967 of these former suburban 'O2' Class 0-4-4Ts introduced by Adams for the LSWR in 1889. Here, No W24 Calbourne (above) and W14 (below) perform mixed duties reminisant of branch line operations half a century ago. The 'O2's were built at Nine Elms to replace the Beattie well-tanks (Page 105) and were a predecessor of the M7s (Page 104).

Wishing Well Bank 1. Happy days on the bank above Weymouth, as BR Standard Class '4', No 75075, attacks the climb with an up parcels.

Above *Wishing Well Bank 2. An hour later, one of the BR Standard '5's, named after a famous 'King Arthur' class engine left Weymouth with an up local. She is No 73118,* King Leodogrance.

Below *A quiet moment at Bournemouth Shed when the yard's only occupant is a BR Standard '3' No 76026, patiently await- ing its next turn of duty on 24 January 1967.*

Above *'Merchant Navy' Pacific No 35008,* Orient Line, *dashes over the lake at Poole with an up local.*

Below *With rhythmic stealth, 'West Country' Pacific No 34034,* Honiton, *heads away from Bournemouth with the 08:35 Express from London (Waterloo) to Weymouth on Monday 29 May 1967.*

I spent countless happy hours on Bournemouth Central Station and on Saturday 24 June 1967 made this study of 'West Country' Pacific No 34021, Dartmoor, approaching from Weymouth with an express from London Waterloo.

Studies in style

Above *Unrebuilt 'Battle of Britain' Pacific No 34057,* Biggin Hill, *nears Winchfield with a Waterloo to Southampton stopping train. A Salisbury engine over the final years,* Biggin Hill *survived until the end of Southern steam and was broken up by Cashmore's of Newport in December 1967.*

Below *Rebuilt 'West Country' Pacific No 34097,* Holsworthy, *races past Worting's grassy banks with a London Waterloo to Bournemouth to Weymouth express.* Holsworthy, *not rebuilt until March 1961, was withdrawn in April 1966 and finally disappeared into Cashmore's Newport yard the following September.*

Chapter 6

London & North Eastern
'A4's

It was a sleepy summer's afternoon at the north end of Grantham Station and some thirty spotters were idly exchanging notes, dreaming of trips to be, or even sitting dozing on a porter's trolley, having been mesmerized by the heat haze rising from the track. In fact, the whole station seemed to be having a siesta; the only sign of life being a sizzling GN 'C12' 4-4-2 tank, whose safety valves periodically sighed desultorily in sleepy indifference as she waited to leave with a stopping train; her departure was long hence, and her crew, nowhere in sight.

Suddenly the home signal lifted for the southbound main, followed immediately by the distant. 'Double Main!' a boy cried, and the group stirred slightly in anticipation. For several minutes, the silence remained, until a faint roar became audible way to the north. But as the roar increased in intensity, a mellow chime whistle rang out its musical tones, lilting yet urgent, 'Streak', and every boy was on his feet, tense with excitement. Again the sonorous cry, a longer blast now, this one for the station, as the sensuous lines of an 'A4' burst out of the distant trees and approached the station at an alarming rate. Thirty pairs of eyes strained for the number; the digits on the front stood proud but uneven; 6002? B? 64B Plate? 60027? It can't be! It is! 60027 64B... a flash of the nameplate — *Merlin*. The throaty roar of three cylinders and scream of passing coaches mixed inextricably with the cheer from the crowd. Notebooks and pens, and even packs of sandwiches, flew into the air as, dancing and backslapping, the tumult lasted long after the train was but a speck far away to the south. *Merlin*, a Haymarket Streak, 'Fantastic'; 'Got her through the main last Saturday'; 'Incredible...'

Left and below *Following dieselization of the East Coast main line, a batch of 'A4' Pacifics was transferred to Scotland in October 1963 for a further lease of life working expresses between Glasgow (Buchanan Street) and Aberdeen. Included was No 60026,* Miles Beevor, *an old favourite from train-spotting days at Grantham when she was a King's Cross engine.*

The 'A4's remained on this service until 1966 although Miles Beevor *was withdrawn in 1965 and stood dumped at her home depot, Aberdeen, Ferryhill, over the following year. In 1967, she was moved to Crewe Works to provide spares for sister No 60010,* Dominion of Canada *which was being preserved for preservation in Montreal. The sad remains of* Miles Beevor *were then taken to Hughes Bolckows yard at North Blyth and broken up in January 1968.*

'A1's and 'A2's

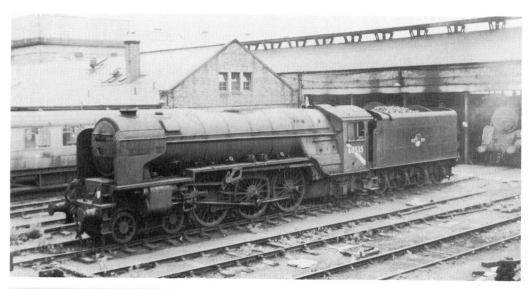

Left *What a challenge the 'A2's posed for trainspotters in the Midlands and South, as many rarely worked the southern reaches of the East Coast main line — especially those allocated to Scotland and Heaton (Newcastle). Also, their propensity to work mail, express parcels and fitted freights meant that much of their running was done at night. A particularly hard one was the engine illustrated here, No 60535,* Hornet's Beauty, *a traditional Edinburgh (Haymarket) engine. One of the last survivors, she ended her days at Glasgow, Polmadie until withdrawal in July 1965. After difficulty, I managed to see all of the 'A2's with the exception of 60539* Bronzino, *a Newcastle (Heaton) engine.*

Below left *'A4's apart, the wonderful family of LNER Pacifics has fared badly in preservation — almost as badly as the celebrated LNWR express passenger classes, although they went before the preservation movement got under way. Of the LNER Pacifics, none of the fifty 'A1's was saved; only one of the forty 'A2's and one of the 78 'A3's! Certainly the 'A1's were around well into the years of preservation, as this picture of No 60124,* Kenilworth *at Darlington in 1965 clearly proves. She lasted until March 1966, and was broken up by Drapers of Hull the following August.*

Below *Another surviving 'A1' was No 60118* Archibald Sturrock, *one of a batch which ended their days at Leeds Neville Hill. She is caught amid the smoky depths of her home depot in April 1965 and, amazingly, still bears her precious nameplates.*

'A3's

Right *If only we had taken pictures then! How many times does one hear this said by those who lived through the years of steam following World War 2. Occasionally — very occasionally — I did take a camera and here is another of my early efforts, depicting an unidentified 'A3' at Grantham.*

Left Salmon Trout, *how we dreamed of her as kids. She was 'off Edinburgh Haymarket' and like her sisters,* Spion Kop *and* Spearmint, *'hadn't worked the main in years'. Redress was made in the final months of* Salmon Trout's *life, as I proudly apply a round of oil at Edinburgh St Margaret's depot in 1965.*

Right *In absolute contrast to the reverence for Haymarket's 'A3's, was the familiar contempt reserved for those on the southern reaches of the East Coast main line. No 60106 Silver Fox, was a Leicester Great Central racehorse for much of the '50s. She returned to the GN in September 1957 and worked from Grantham to Peterborough New England until withdrawal in December 1964. She is seen forlornly reposing beneath the abandoned coaling plant at Peterborough New England shed around Christmas 1964, waiting to go to King's of Norwich, where she was broken up the following year.*

Left *The 'A3's were true thoroughbreds, like the famous Doncaster-winning racehorses they were named after. But their beauty was marred during the late '50s, when double-chimneys were fitted, followed in the early '60s by German style smoke deflectors, as witness this study of Salmon Trout. No other British design received such smoke deflectors, and although the result was imposing, the 'A3's traditional beauty was inevitably spoiled.*

The first 'A3' to be withdrawn was the old Leicester engine No 60104, Solario, in 1959. The next withdrawals took place during 1961, and I remember visiting Doncaster Works in October that year and seeing No 60055, Woolwinder, and 60064, Tagalie, in the process of being cut up; the shock was tremendous as, throughout our adolescence, we knew the 'A3's would last forever.

Scottish inside cylinder 0-6-0s

Above *0-6-0s didn't carry names! They were far too humble amid the hierarchy of mainline designs for such flamboyance. But a delightful exception occurred with the North British Railway's classic 'J36's, 25 of which saw overseas service during World War 1 and upon return were given commemorative names;* Mons, Plumer, Verdun, Ypres, *are four which immediately come to mind, along with* Byng *which I saw on Carlisle Canal Shed in 1955. I caught this unnamed survivor backing into Thornton Junction shed in 1965, only weeks before she was withdrawn.*

Above left *During the Edwardian period, Britain's 0-6-0s became increasingly potent in the never-ending demand for more power. This trend led the North British Railway to augment its Standard 'J36's, first with the 'J35's and, in 1914, with these even bigger, superheated 'J37's. Over one hundred were put into service between 1914/21.*

Left *J37s could be seen at work on the Fifeshire coalfield until as late as 1965, providing a delightful pre-grouping atmosphere to a scene which was becoming increasingly standardized.*

The inside-cylinder 0-6-0 was the backbone of British motive power; 'The Maid of all Work' and true British drudge. Built continuously for over a century, countless delightful variations were made upon the theme; many were essentially identical but generally getting larger as the years passed. In 1913, over 7,200 were active — 46% of the total mainline stock whilst, at nationalization in 1948, they still accounted for 37% of the total!

I have always revered the 0-6-0 and taken them very seriously. Never forget a point made by Dr Tuplin when he recounted the story of a high-ranking railway official giving a lecture on the LNER Streamlined trains, in which he described them as 'Chief Mechanical Engineer's toys', and went on to say that it was the 0-6-0s which earned the money.

North Eastern inside cylinder 0-6-0s

Above left and above *Blyth, a last stronghold of the former North Eastern Railway's superb 'J27's, the last in their long progression of ever more powerful 0-6-0s. A total of 115 were built between 1906/22. As late as 1965 some fifty examples remained active hauling coal trains over their original routes in north-east England.*

Freight sheds on a Sunday afternoon possessed a captivating tranquility. Walking beside the still giants, every footstep reverberated around the depot, and if one crushed a piece of clinker underfoot, the sound was almost deafening. A heady aroma of soot pervaded one's senses. It was a particularly acrid, cold smell, so suggestive of latent power. It was on such occasions that one got really close to the engines and noticed their detailed differences, for when they were up and about, the exuberant vitality and all-pervading life made pensive reflection difficult. I spent many Sunday afternoons wandering through a shed of twenty engines, like this occasion at Blyth: 'cabbing' them, handling regulators, opening fire-hole doors and pulling silent whistle cords.

These were times when you wrote down works plate details and searched for tell-tale signs of old numbers from former companies, still proud to the searching eye, beneath an generation's over-painting. Inevitably, blackened hands smeared any notes made, and by the time you left the shed, the grime of your pleasure would have smudged both cheeks.

Eight-coupled 'Q6's

The former North Eastern Railway 'Q6's were the last surviving examples of the 0-8-0 freight engine, which some railways chose as a natural progression from the ubiquitous 0-6-0. An obvious example of this was the LNWR, although other railways — like the Great Western — passed directly from 0-6-0s to 2-8-0s.

The NER, with its vast coalfield traffic, remained content with 0-8-0s, the 'Q6' Class being comprised of 120 engines built between 1913 and 1921. They proved perfect for the NE's slow slogging coal hauls. In 1919, a three-cylinder version was introduced, Class 'Q7', but these were withdrawn *en-bloc* in 1962.

A shedman shovels clinker from the smoke-box of a 'Q6' at West Hartlepool in 1965.

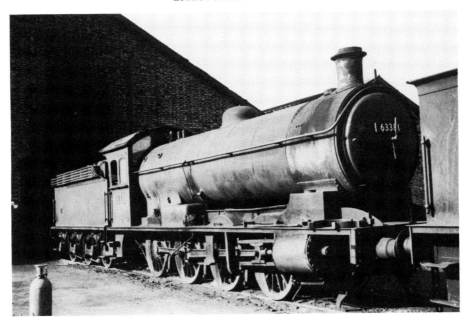

Above *The 'Q6's survived intact until 1960, and the last examples were not withdrawn from the North-east until 1967. Old as they were, they had many more years of life left within them.*

Below *It was marvellous to see an ex-works 'Q6' in 1965. She must have been one of the last engines to be outshopped from Darlington, as this once great North Eastern Railway Works closed early that year.*

Eight-coupled: general

Above *A superb contrast in motive power at West Hartlepool featuring a WD 'Austerity' 2-8-0, alongside a 'Q6' 0-8-0. The North Eastern Railway, despite its vast coal traffic, never advanced beyond the 0-8-0, obviously feeling that, for sustained heavy pulling over good tracks, the type offered maximum adhesion. The influx of WD 2-8-0s after the Second World War enhanced the coalfield services and provided a considerable increase in power and yet there were many North Eastern crews who claimed they could get better work from their traditional 'Q6's.*

Above left *My favourite spotting place at Rugby was at the girder bridge where the Great Central crossed the West Coast main line. Coal traffic heading south along the Central from Nottinghamshire was prolific, and how these 'Annesley Rods' (as we called them) used to thunder across those girders with their long, heavy trains. There were many varieties, all various re-builds of the original Robinson 2-8-0s (below) and this one has been re-built with a Thompson B1 type boiler and side-window cab, but has retained her original cylinders, frames and splashers.*

Left *The handsome lines of a Robinson Great Central 2-8-0 caught in 1965 in her original unmodified form. A classic amongst 2-8-0s, the type was introduced in 1911 as the Great Central's heavy mineral engine, but within three years had also been adopted by the Railway Operating Division for overseas service during the First World War.*

The British freight locomotive reached its maturity with the 2-8-0, and it was only at the eleventh hour that it was partly superseded by BR's Standard '9F' 2-10-0s. For Britain, the 2-8-0 was a twentieth century concept, and although it began as early as 1903, with Churchward's '2800's for the Great Western, it took many years for the type to achieve any prevalence and replace the 0-8-0 — of which it was the logical development — but more importantly, the inside-cylinder 0-6-0. By 1914, only 206 2-8-0s were in national service, and even in 1948, there were little more than 1,500 as against almost 4,500 0-6-0s.

'Austerities': 1

No class was so unkempt and scruffy as these 'Austerity' 2-8-0s. They were the filthiest engines ever to run in Britain and if one was seen looking clean, it was certain to be an ex-shopper. The famous 'Willesden Grey', which many Stanier '8F's suffered from, was a mere coating of dust compared with the layers of embedded grime and grease which insulated every 'Austerity' boiler. Perhaps they were unloved because they were anybody's engine; they had no true lineage and, following the war, were literally drafted nationwide.

Their stubby, cast-iron chimneys, pre-fabricated parts and the 'plonk — bang — plonk — bang — plonk — bang' of their motion/bushes whenever they moved inspired no-one. Though abused in every respect, they were superb pulling engines; perfect for wartime and perfect for the rough and tumble afterwards.

For all serious trainspotters, they were perhaps the most exciting class in Britain; no type crossed regional boundaries like the 'Austerities.' They anticipated the free movement of diesel locomotives twelve years hence and, as the 733 members of the class were spread between 66 different depots, all were potential rarities.

Sitting by my bridge at Newton, I always knew when one was approaching by the banging motion. They were like a try in rugby-football, a 50 per cent chance of scoring. It could be off Shrewbury, Mexborough, Hull or wherever, but the problem was reading the number and on several occasions, examples passed bearing a rare shed code, but unidentified! As an example of what fair game they were, I ended up needing 136 of them, despite fifteen years of avid trainspotting throughout the country!

Their pugnacity and utility ensured that some survived virtually until the end of steam, yet not one was preserved — as I said earlier, they were anybody's engine and devoid of lineage. A decade later, in more enlightened times, one was found stored in a tunnel in Sweden, and this has been returned to Britain, as testimony to one of the most fascinating locomotive designs in the history of the steam locomotive.

Once steam had ended on the Southern our attentions were focused on the north of England and, on Saturday 16 September 1967, I began an extensive two-week tour. Sadly, action in the North-East was over, but the engines remained and here 'Austerity' No 90135 reposes beyond the smashed windows of Sunderland Shed on 21 September.

Sunday afternoon and 'Austerities' galore at Normanton Shed. Nos 90615 and 90722 sit in silent repose behind two sisters on 17 September 1967.

'Austerities': 2

Left *Six somnolescent 'Austerities' pass a quiet Sunday in Doncaster shed yard. Introduced in 1943, the 'Austerities' were a much simplified and 'fabricated' version of the Stanier '8F'. They were meant to give several years' hard, unstinted service primarily for the War Department overseas. After nationalization an incredible 733 of them became integrated into British Railway's stock.*

Below left *Wherever there was heavy hauling to do, there were 'Austerities', and the Great Central section abounded with them. This example caught marshalling a coal train at Staveley is so characteristically grimy that her number was totally indistinguishable.*

Above right *An 'Austerity' in pure form! A Stanier '8F' returned from overseas war service, at Derby Works on 30 August 1950. It was my first visit to Derby Works and I remember how this war veteran from the Middle East fascinated me. Note the Johnson Belpair 4-4-0 in the background.*

Middle right *'Austerity' No 90041 was one of many examples allocated to Scotland, working from such sheds as Aberdeen Ferry Hill, Thornton Junction, Dundee, Dunfermline, Eastfield and Hamilton.*

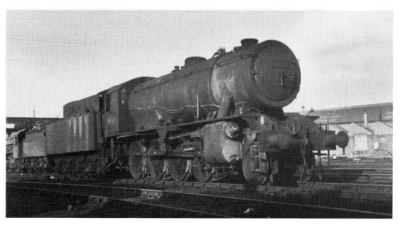

Right *The stolid, inelegant line of the 'Austerity' is captured to perfection in this study of No 90010.*

Inside cylinder 4-4-0s

Above *I was on one of my regular visits to Stratford Shed in London, said to have the biggest allocation in the country. In addition to the vast shed was Polygon Works which was never easy to 'bunk', yet always worth the effort. Especially on this occasion — it was 29 September 1957 and I found the celebrated 'D16' 4-4-0, Claud Hamilton, up from Yarmouth (South Town) depot for scrap.*

The original Claud Hamilton was the first of a celebrated class of Great Eastern Passenger Express 4-4-0s which emerged from Stratford Works in 1900. During the '30s, Claud Hamilton was rebuilt with a larger boiler and modified framing and, following her withdrawal from service in 1947, the nameplates were transferred to sister engine No 62546.

Above left *From an early age I dreamed of seeing the great North of Scotland Railway 'D40' 4-4-0s. Their beauty captivated me and, by the age of twelve, I wondered however I might get to see them in their far-away Speyside haunts. Their names were steeped in romance; Benachie; Gordon Highlander; Glen Grant.*

My chance came in 1955 when my father organized a cycle tour of Scotland and, accompanied by Brian Stafford, we travelled part train and part cycle through the wonders of Scotland's railway network, fully embracing the industrial intrigue of Ayrshire, Glasgow, Edinburgh and Fifeshire. But the crowning glory was in the far North with the resplendant 'D40's — even more beautiful than I had imagined. Here is No 62265 at Boat of Garten, the farthest point west on the former GN of S system. On the extreme right can be seen the front of former Highland Railway inside cylinder 4-4-0, Ben Alder.

Left *Our visit to Inverurie, the former GN of S Railway Works, produced examples of a very different 4-4-0. Also of noble lineage, yet lacking the Victorian grace of a 'D40', for No 62692, Alan Bane was one of the Scottish 'Directors', built during the early 1920s to the design of Robinson's Great Central directors. There were 24 of these locomotives spread between Glasgow, Eastfield and Edinburgh, Haymarket depots, and this engine had obviously come to Inverurie for shopping. Notice the 'LNER', still visible under the 'BR' emblem on her tender.*

Their names were amongst the most inspiring in British locomotive history. All came from Walter Scott's novels: Bailie MacWheeble; Luckie Mucklebackit; Wizard of the Moor; Laird of Balmawhapple along with many others, and over my various visits to Scotland, I saw all of them. Withdrawal commenced in 1958 and they had all disappeared by 1962.

Victorian designers made great works of art of their inside cylinder 4-4-0s; it was a wheel arrangement of elegance and symmetry which, combined with the aesthetical prowess of appearance, produced some beautiful locomotives.

Delightful to the eye, delightful to the ear. Never will I forget the lovely musical plonk of coupled wheels as the former Midland Railway 'Simples' rounded the curve at Wigston North Junction.

LNE Round-topped

Left *The all-purpose 'K1' 2-6-0, introduced in 1949 by A. H. Peppercorn, was intended to replace older pre-grouping classes on a wide variety of duties: 71 were put into service and they remained intact until 1962.*

Middle left *Spotters at Leicester London Road watch the approach of the mid-day Norwich to Birmingham buffet car train, which brought a whiff of East Anglian magic to the Midlands scene, usually in the form of an ex-Great Eastern 'B12' 4-6-0; a re-built Claud 4-4-0 — as on this occasion — or, disappointingly, an Ivatt '3000' mogul from South Lynn (31D) depot. The visitor used to detach at Leicester Station and stand 'middle line' with its pre-grouping buffet car, before working back during mid-afternoon.*

Below *Named 'B1's were always exciting — there were only 59 of them from a total of 410 engines. Forty were named after African antelopes, eighteen after LNER directors and one,* Mayflower. *Here is No 61014,* Oribi, *a truly rare beast from Tweedmouth (52D).*

Above *Cohen's scrapyard at Cransley near Kettering received an amazing variety of locomotives and in this unhappy scene, a brace of LNER 'B1's from Doncaster are in the process of being broken up.*

Right *We were off to Swindon Works on a loco spotter's special; it was April 1954 and I was so excited, I hardly slept a wink the night before. Our train began in Nottingham and was headed by No 61283, a Colwick (38A) 'B1'. I made this picture before we joined the train at Leicester Central. It was a fabulous day as, in 1954, Swindon remained immersed in Great Western traditions (although we did also see some BR Standard '3' Moguls being built).*

Right *Leicester Great Northern Shed was a regular part of our itinerary as we cycled round the city's sheds. There was never much on, usually a couple of ex-GN 'J5' or 'J6' class 0-6-0s, but occasionally there would be a 'Ragtimer K2' from Colwick — as on this sunny afternoon in 1951, when I found No 61777 simmering quietly in the yard.*

The incredible 'J72's

By an amazing quirk of evolution, these unassuming little tank engines were built virtually unaltered over a period of 54 years! No other British design achieved such distinction, which is possibly a record in world history.

When I first saw 'J72's, decked in green livery for pilot duty at York and Newcastle Stations I could hardly believe that they had just been built; an old design of pre-grouping tank engine, being built by British Railways in 1951 — incredible! At that time, diesels were accepted as the shunting engines of the future and whatever gaps needed to be filled in the interim could easily have been covered by the many ageing 0-6-0s or smaller pre-grouping tanks of semi-redundant status.

The story began in 1898 when Wilson Worsdell, the Chief Mechanical Engineer of the North Eastern Railway, produced the first twenty engines at Darlington Works. A further twenty appeared in 1914, during Sir Vincent Raven's tenure as CME. Ten followed in 1920 and another 25 two years later. After the grouping, Nigel Gresley became CME of the newly-formed LNER and one of his first actions was to build another ten J72s at Doncaster in 1925!

With 85 engines in traffic, the chapter was surely closed, and so it was for the next quarter century until,

in 1950, following the nationalization, British Railways suddenly ordered another 28 examples from Darlington Works, the last of which was completed in 1951. The class remained intact until 1958, when some of the original 1898 batch were withdrawn. Not surprisingly, the type lasted almost until the end of steam — a remarkable and practical tribute indeed to the soundness of their basic design.

Below left and below.*A pair of ex-North Eastern Railway 'J72' 0-6-0Ts end their days in departmental service on the North Eastern region.*

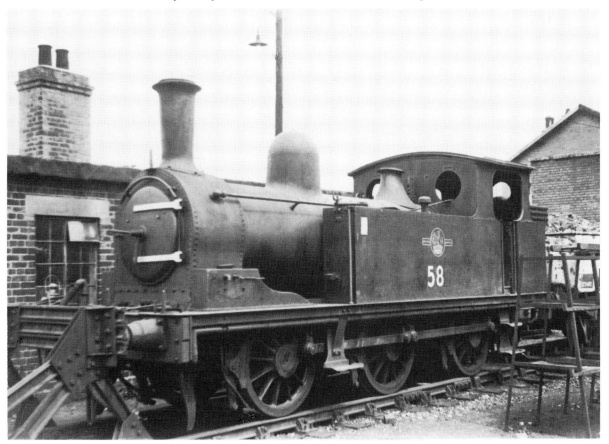

Bon Accord

Left, below and right *LNER 'A1' Pacific No 60154* Bon Accord, *during her final months in service at Leeds (Neville Hill). She was one of a class of fifty engines which worked expresses on the East Coast main line, and throughout the 1950s, the type could be seen anywhere between King's Cross and Aberdeen.*

Bon Accord was built under British Railways at Doncaster in 1949 as one of a batch of five 'A1's fitted with roller bearings, which enabled them to achieve very high mileages between overhauls. She was allocated to Gateshead (52A) as one of their fourteen 'A1's. For us in the Midlands, Gateshead was one of those potentially difficult depots whose express engines could either work north or south, as opposed to Haymarket's 'A1's, which had no regular workings to London, or in contrast, King's Cross and Peterborough's examples which could only work north.

For me, *Bon Accord* was one of the commoner Gateshead 'A1's on the southern reaches of the East Coast main line, and I saw her many times, whereas No 60142 *Edward Fletcher* — another long-term Gateshead engine — was very hard; she was my last 'A1', and I never did see her in the south.

There used to be a theory — probably unfounded — that depots like Gateshead, which had a choice of direction, would use certain engines to work north, and others to work south, and seldom was the allocation operated at random. Certainly, Gateshead's *Meg Merrillies* and *Kenilworth* seemed common at Grantham, whilst their *Redgauntlet, Willbrook* and *Borderer* were not.

In November 1960, dieselization caused *Bon Accord* to be transferred from Gateshead to York, where she remained until going to Leeds (Neville Hill) in July 1963. Her final years were spent on duties which became increasingly general as diesels took over the more prestigious jobs. The accompanying pictures were made on a special trip to Leeds on 23 April 1965.

She was finally withdrawn the following September, a superb Pacific, only sixteen years old. She stood dumped at Neville Hill until November, when, along with sister engines No 60118, *Archibald Sturrock* and 60134, *Foxhunter,* she was towed to Sheffield. They were taken to Ward's Scrapyard at Beighton and, by the end of December, all three had been broken up.

Scottish 'A2's

The 'A2's had a mystique which for me was epitomised by an amazing track on the Argo record 'Rhythms of Steam' featuring *Blue Peter* heading southwards from Aberdeen towards Lunan Bay with the Aberdonian sleeping car express. The rapid purring of her soft staccato exhaust beats, syncopated by three-cylinder rhythms, could be heard several miles away, and the next few minutes are sheer bliss as the hypnotic, foot-tapping sound became ever closer until, with a climactic rasping swish, the Pacific passed the microphone and, after the brief drumming sound of her ten-coach train, the engine's exhaust became clear again and upon receding, intermittently acquired deeper overtones whenever the line passed through slightly raised embankments.

Whether in sound, history, design variations, or simply as rare birds, the LNER 'A2's were one of our most inspiring classes, not least as the first six were Thompson's 1943 rebuilds of Gresley's magnificent Mikados, specially designed for the difficult Edinburgh to Aberdeen route. The rebuilds took the old Mikado names.

The following year, four more 'A2's appeared with 'V2' 2-6-2 type boilers. A further batch appeared in 1946, based on the original rebuilds, and during 1947-8, Thompson's successor, A. H. Peppercorn, produced a further series which brought the class total to forty. Exotic names added much to the allure of rare engines and for most spotters some 'A2's were rare. Let us stop the clock at 1955 — a year of fine vintage for steam — and see how the 'A2's were distributed. The table below summarizes shed allocations in that year.

The accompanying pictures were made on my Scottish tour of 1965, which constituted a farewell as the last three survivors, *Tudor Minstrel, Sayajirao* and *Blue Peter,* were all at Dundee Tay Bridge. It was especially sad as, a few days earlier, I had seen the last 'A3' Pacifics at Edinburgh St Margaret's.

'A2' Pacific No 60530, Sayajirao — an Edinburgh Haymarket engine for many years — ends her days at Dundee, Tay Bridge. Withdrawn in November 1966, she was broken up the following March by the Motherwell Machinery and Scrap Company at Wishaw.

Shed allocations of 'A2's in 1955

60500	*Edward Thompson*	Peterborough (New England)	35A	60520	*Owen Tudor*	Peterborough (New England)	35A
60501	*Cock o' the North*	York	50A	60521	*Watling Street*	Gateshead	52A
60502	*Earl Marischal*	York	50A	60522	*Straight Deal*	York	50A
60503	*Lord President*	York	50A	60523	*Sun Castle*	Peterborough (New England)	35A
60504	*Mons Meg*	Peterborough (New England)	35A	60524	*Herringbone*	York	50A
60505	*Thane of Fife*	Peterborough (New England)	35A	60525	*A.H. Peppercorn*	Aberdeen (Ferryhill)	61B
60506	*Wolf of Badenoch*	Peterborough (New England)	35A	60526	*Sugar Palm*	York	50A
60507	*Highland Chieftain*	Edinburgh (Haymarket)	64B	60527	*Sun Chariot*	Dundee (Tay Bridge)	62B
60508	*Duke of Rothesay*	Peterborough (New England)	35A	60528	*Tudor Minstrel*	Dundee (Tay Bridge)	62B
60509	*Waverley*	Edinburgh (Haymarket)	64B	60529	*Pearl Diver*	Edinburgh (Haymarket)	64B
60510	*Robert the Bruce*	Edinburgh (Haymarket)	64B	60530	*Sayajirao*	Edinburgh (Haymarket)	64B
60511	*Airborne*	Newcastle (Heaton)	52B	60531	*Bahram*	Aberdeen (Ferryhill)	61B
60512	*Steady Aim*	York	50A	60532	*Blue Peter*	Aberdeen (Ferryhill)	61B
60513	*Dante*	Peterborough (New England)	35A	60533	*Happy Knight*	Grantham	35B
60514	*Chamossaire*	Peterborough (New England)	35A	60534	*Irish Elegance*	Edinburgh (Haymarket)	64B
60515	*Sun Stream*	York	50A	60535	*Hornet's Beauty*	Edinburgh (Haymarket)	64B
60516	*Hycilla*	Gateshead	52A	60536	*Trimbush*	Edinburgh (Haymarket)	64B
60517	*Ocean Swell*	Newcastle (Heaton)	52B	60537	*Bachelor's Button*	Edinburgh (Haymarket)	64B
60518	*Tehran*	Gateshead	52A	60538	*Velocity*	Gateshead	52A
60519	*Honeyway*	Edinburgh (Haymarket)	64B	60539	*Bronzino*	Newcastle (Heaton)	52B

Blue Peter, *built at Doncaster in 1948, was one of the 'A2' stud allocated to Aberdeen, Ferry Hill. She moved to Dundee Tay Bridge in June 1961 until her withdrawal in December, 1966. Happily preserved, she is the sole representative of the LNER's Thompson and Peppercorn Pacifics, embraced by classes 'A1'/'A2'.*

Index

Great Western

Ex-Rhymney	0-6-2T	*22*
Ex-Taff Vale	0-6-2T	*22, 86*
'1400'	0-4-2T	*8, 9*
'2021'	0-6-0PT	*14, 15*
Ex-BPGV	0-6-0T	*23*
'2800'	2-8-0	*20, 21, 90, 127*
'Saint'	4-6-0	*17*
'Bulldog'	4-4-0	*17*
'5700'	0-6-0PT	*14, 15*
'Castle'	4-6-0	*10, 11, 12, 13, 24, 29*
'4100'	2-6-2T	*9*
'Hall'	4-6-0	*16, 17, 18, 19, 24, 67*
'Grange'	4-6-0	*24, 67*

London Midland & Scottish

Fowler	2-6-2T	*51*
Ex-MR 'Simple'	4-4-0	*26, 133*
Ex-MR 'Belpair'	4-4-0	*131*
Compound	4-4-0	*58, 59*
Ivatt	2-6-2T	*27, 92*
Ex-MR '0F'	0-6-0T	*55*
Ex-MR '1F'	0-6-0T	*7, 54, 55, 131*
Fairburn	2-6-4T	*37, 52, 53, 92*
Fowler '2300'	2-6-4T	*50, 51, 53*
Stanier	2-6-4T	*92*
'Crab'	2-6-0	*40, 44, 45, 47*
Stanier 'Crab'	2-6-0	*46, 47*
Ivatt '3000'	2-6-0	*49, 134*
Ex-MR '3F'	0-6-0	*60, 61*
'4F'	0-6-0	*55, 63, 82*
'Black 5'	4-6-0	*29, 30, 31, 34, 35, 37, 40, 56, 57, 58, 65, 66, 67*
'Jubilee'	4-6-0	*7, 17, 26, 30, 35, 38, 39, 40, 41, 47, 58*
'Royal Scot'	4-6-0	*41, 42, 43*
'Princess Royal'	4-6-2	*89*
Turbomotive	4-6-2	*89*
Ivatt '6400'	2-6-0	*48, 49*
Ex-MR 'Jinty'	0-6-0T	*54*
Stanier '8F'	2-8-0	*Front endpaper, 20, 29, 33, 36, 64, 65, 67, 90, 128, 131*
Ex-LNWR	0-8-0	*62, 63*
Ex-Caledonian	4-4-0	*58, 59*
Ex-Caledonian	0-6-0T	*32*
Ex-MR '1P'	0-4-4T	*27*
Ex-MR '2F'	0-6-0	*60, 61, 92*

Industrial

Andrew Barclay	0-4-0ST	*72, 73, 78, 79*
Andrew Barclay	0-6-0ST	*84*
Avonside	0-6-0ST	*80, 81*
Bagnall	0-6-0ST	*73*
Hawthorn Leslie	0-4-0ST	*74*
Hudswell Clarke	0-4-0ST	*74*

Hudswell Clarke	0-6-0T	*82, 83*
Hunslet	0-6-0ST	*80*
Hunslet 'Austerity'	0-6-0ST	*70, 74, 75, 84, 85*
Kitson	0-6-0ST	*71*
Kitson	0-6-2T	*86, 87*
Lambton	0-6-0	*86*
Mersey Railway	0-6-4T	*77*
Robert Stephenson & Hawthorn	0-4-0CT	*68, 69*
Robert Stephenson & Hawthorn	0-4-0ST	*76*
Robert Stephenson & Hawthorn	0-6-0ST	*70, 74, 78, 82*
Robert Stephenson	0-6-2T	*86, 87*

British Railways

'Britannia'	4-6-2	*88, 89*
Duke of Gloucester	4-6-2	*89*
'Clan'	4-6-2	*88*
'73000'	4-6-0	*27, 30, 94, 96, 97, 103, 110*
'75000'	4-6-0	*94, 95, 109*
'76000'	2-6-0	*49, 95, 110*
'77000'	2-6-0	*95*
'78XXX'	2-6-0	*Frontispiece, 49, 92*
'80000'	2-6-4T	*53, 92*
'82000'	2-6-2T	*96*
'84000'	2-6-2T	*92*
'9F'	2-10-0	*20, 24, 90, 91, 127*

Southern

Ex-LSWR 'M7'	0-4-4T	*104, 105, 108*
Ex-USATC	0-6-0T	*104, 106, 107*
Ex-LSWR 'B4'	0-4-0T	*104, 106*
Ex-LSWR 'O2'	0-4-4T	*108*
Ex-LSWR 'Beattie'	2-4-0WT	*105, 108*
'King Arthur'	4-6-0	*103, 110*
Ex-PDSWR	0-6-2T	*105*
'S15'	4-6-0	*102, 103*
'N'	2-6-0	*102, 103*
'West Country'	4-6-2	*98, 99, 111, 112, 113*
'Battle of Britain'	4-6-2	*98, 99, 100*
'Merchant Navy'	4-6-2	*94, 98, 99, 100, 101, 111*

London and North Eastern

'A4'	4-6-2	*114, 115*
'A3'	4-6-2	*67, 118, 119, 140*
'A1'	4-6-2	*116, 117, 138, 139, 141*
'P2'	2-8-2	*140*
'A2'	4-6-2	*Rear endpaper, 116, 117, 140, 141*
'V2'	2-6-2	*140*
'B1'	4-6-0	*134, 135*
'B12'	4-6-0	*134*
'K2'	2-6-0	*135*
'K1'	2-6-0	*134*
Ex-GNSR 'D40'	4-4-0	*132, 133*
Claud	4-4-0	*133, 134*
'D11'	4-4-0	*132, 133*
Ex-NE 'Q6'	0-8-0	*124, 125, 127*

Ex-NE 'Q7'	0-8-0	*124*
Ex-GC 'O4'	2-8-0	*126, 127*
'O4/8'	2-8-0	*126, 127*
Ex-GN 'J6'	0-6-0	*135*
Ex-NE 'J35'	0-6-0	*121*
Ex-NB 'J37'	0-6-0	*120, 121*
Ex-NB 'J36'	0-6-0	*121*
Ex-GN 'J5'	0-6-0	*135*
Ex-NE 'J27'	0-6-0	*122, 123*
Ex-GN 'C12'	4-4-2T	*115*
'Austerity'	2-8-0	*127, 128, 129, 130, 131*

'A2' Pacific No 60530, Sayajirao —
an Edinburgh Haymarket engine.